狡猾的思想

The Network of Thought

［印］克里希那穆提　著

张春城　译

北京时代华文书局

CONTENTS

目录

第一章

意识的真实面貌

Truth of Consciousness

>> I see some of my old friends are here—and I am glad to see you. As we are going to have seven talks we should go into what I am going to say very carefully, covering the whole field of life, so please be patient those of you who have heard the speaker before, please be tolerant if the speaker repeats himself, for repetition has a certain value.

我看到我的一些老朋友在这里——很高兴看到你们。我们会进行七次谈话，将非常认真地探究我要说的话，其涵盖生活的整个领域，因此请那些以前听过我讲话的人保持耐心，容忍讲话者重复他曾讲过的话，因为重复的内容具有某种价值。

Prejudice has something in common with ideals, beliefs and faiths. We must be able to think together; but our prejudices, our ideals and so on, limit the capacity and the energy required to think, to observe and examine together so as to discover for ourselves what lies behind all the confusion, misery, terror, destruction and tremendous violence in the world. To understand, not only the mere outward facts that are taking place, but also the depth and the significance of all this, we must be able to observe together—not you observing one way and the speaker another, but together observe the same thing. That observation, that examination, is prevented if we cling to our prejudices, to our particular experiences and our particular comprehension. Thinking together is tremendously important. Where nothing is sacred, where no one respects another. To understand all this, not only superficially, casually, we have to enter into the depths of it, into what lies behind it. We have to enquire why it is that after all these millions of years of evolution, man, you and the whole world, have become so violent, callous, destructive, enduring wars and the atomic bomb. The technological world is evolving more and more; perhaps that may be one of the factors causing man to become like this. So, please let us think together, not according to my way or your way, but simply using the capacity to think.

成见和理想、信念、信仰具有某种共通性。我们必须一同思考，为了亲自发现所有困惑、不幸、恐怖、破坏以及这个世界上惊人的暴力背后的东西，我们需要共同观察和检视，但是我们的成见、理想等限制了这样做所需要的能力和能量。不仅要了解正在讨论的表面事实，还要了解这一切的深刻内涵和意义，我们必须能够共同观察——不是你观察你的，讲话者观察他自己的，而是共同观察相同的东西。如果我们坚持我们的成见、特定的经验和理解，那份观察、那份检视就被阻碍了。共同思考极其重要。假如神不存在了，人们将不再尊重彼此。要了解所有这一切，不是肤浅地、随意地了解，我们必须深入探究它，探究它背后的东西。我们必须探究为什么在经过数百万年的进化之后，我们人类会变得如此暴力、残酷、具有毁灭性，且饱受战争和原子弹的威胁。科技的发展日新月异，这也许是人类变成今天这样的原因之一。因此，请让我们一同来思考，不是按照我的或你的方式思考，而是单纯地运用思考的能力。

>> Thought is the common factor of all mankind. There is no Eastern thought, or Western thought; there is only the common capacity to think, whether one is utterly poor or most sophisticated, living in an affluent society. Whether a surgeon, a carpenter, a labourer in the field, or a great poet, thought is the common factor of all of us. We do not seem to realize that thought is the common factor that binds us all. You think according to your capacity, your energy, your experience and knowledge; another thinks differently according to his experience and conditioning. We are all caught in this network of thought. This is a fact, indisputable and actual.

思想是整个人类的共同要素。不存在东方的思想或西方的思想，只有共同的思考能力，无论你涉世未深还是成熟老练；无论是外科医生、木匠、在田里劳动的人，还是伟大的诗人，思想都是我们所有人的共同要素。但我们似乎没有意识到，思想也是困住我们所有人的共同要素。你基于你的才能、精力、经验和知识来思考，另一个人基于他的经验和训练进行不同的思考，所以我们都陷在这张思想的网中。这是一个不争的事实。

>> We have been "programmed" biologically, physically and also "programmed" mentally, intellectually. We must be aware of having been programmed, like a computer. Computers are programmed by experts to produce the results that they want. And these computers will outstrip man in thought. These computers can gather experience, and from that experience learn, accumulate knowledge, according to their programme. Gradually they are going to outstrip all our thinking in accuracy and with greater speed.

Of course they cannot compose as Beethoven, or as Keats, but they will outstrip our thinking.

我们已经在生物学意义上、在生理上都被“编程”了，在精神上和心智上也是如此。我们必须意识到这一点，就像计算机一样。专家为了得到他们想要的结果，设定了计算机程序。而这些计算机将在思想上胜过人类。根据它们的程序，这些计算机能够收集经验，并从经验中学习和累积知识。逐渐地，它们将在精确性方面以越来越快的速度超越我们所有的思维。当然，它们无法像贝多芬或济慈那样进行创作，但是它们将在思维上胜过我们。

>> So, then, what is man? He has been programmed to be Catholic, Protestant, to be Italian or British and so on. For centuries he has been programmed—to believe, to have faith, to follow certain rituals, certain dogmas; programmed to be nationalistic and to go to war. So his brain has become as a computer but not so capable because his thought is limited, whereas the computer, although being also limited,

is able to think much more rapidly than the human being and can outstrip him.

那么，人是什么？他被设定为天主教徒、新教徒、意大利人、英国人，等等。很多个世纪以来，他被设定成宗教主义者去相信、信仰、追随特定的仪式和教条，被设定成民族主义者去参加战争。于是他的脑子成了一台计算机，但没那么能干，因为他的思想是局限的，而计算机尽管也是局限的，却能够比人更迅速地思考，因此能够超过人类。

>> These are facts, this is what actually is going on. Then what becomes of man? Then what is man? If the robots and the computer can do almost all that the human being can do, then what is the future society of man? When cars can be built by the robot and the computer—probably much better—then what is going to become of man as a social entity? We're facing these and many other problems. You cannot any more think as Christians, Buddhists, Hindus and

Muslims. We are facing a tremendous crisis; a crisis which the politicians can never solve because they are programmed to think in a particular way—nor can the scientists understand or solve the crisis; nor yet the business world, the world of money. The turning point, the perceptive decision, the challenge, is not in politics, in religion, in the scientific world, it is in our consciousness. One has to understand the consciousness of mankind, which has brought us to this point. One has to be very serious about this matter because we are really facing something very dangerous in the world—where there is the proliferation of the atomic bomb which some lunatics will turn on. We all must be aware of all this.

这些都是事实，正在实际发生着。那么人成了什么？人是什么？如果几乎一切人做的事情都能被机器人和计算机代替，那么将来的人类社会将会是什么样子？当机器人和计算机能够制造汽车——也许制造得更好——那么作为一种社会性的存在，人会成为什么？

我们正在面临这些问题，以及很多其他的问题。你不能再作为基督徒、佛教徒、印度教徒和穆斯林来思考了。我们正面临一个巨大的危机——一个政治家不能解决的危机，因为他们习惯于用一种特殊的方式思考；科学家也不能，工商界、金融界的人也不能。那个转折点，那个决断，那个挑战，不在政治、宗教、科学的世界，它在我们的意识中。你必须理解人类的意识，是它将我们带到了这里。对待这个问题，你必须非常严肃，因为我们的世界真的在面临严重的威胁——原子弹激增，而一些疯狂的人会启动它。我们都必须意识到这一切。

>> One has to be very very serious, not flippant, not casual but concerned, to understand this behaviour and how human thought has brought us all to this point. We must be able to penetrate very carefully, hesitantly, with deep observation, to understand together what is happening both out there and inwardly. The inward psychological activity always overcomes the outer, however many regulations, sanctions,

decisions you may have outwardly, all these are shattered by our psychological desires, fears and anxieties, by the longing for security. Unless we understand that, whatever outward semblance of order we may have, inward disorder always overcomes that which is outwardly conforming, disciplined, regularized. There may be carefully constructed institutions—political, religious, economic—but whatever the construction of these may be, unless our inward consciousness is in total order, inward disorder will always overcome the outer. We have seen this historically, it is happening now in front of our eyes. This is a fact.

你必须非常非常认真地理解这些行为，以及人类的思想是如何把我们带到了这里的，而不是草率地、偶尔才关心一下。我们必须能够深入地观察，非常仔细且带着质询之心探究，一同去了解那些外在和内在正在发生的事情。内在的、心理上的活动总是能克服外在影响，无论你有多少外在的规则、处罚、决议，所有这些都被我们心理上的欲望、恐惧、焦虑，以及对安全的渴望打碎了。除非我们理解它们，否则无论我们外在拥有什么样的秩序，内在的混乱总会压倒外在的那些遵从、纪律、规定。或许存在周密构建的机制——政治的、宗教的、经济的——但是无论存在什么样的机制，除非我们内心的意识处于完整的秩序之中，否则内在的混乱就总会压倒外在。我们从历史上看到过这些，它现在正在我们的眼前发生。这是一个事实。

>> The turning point is in our consciousness. Our consciousness is a very complicated affair. Volumes have been written about it, both in the East and in the West. We are not aware of our own consciousness; to examine that consciousness in all its complexity one has to be free to look, to be choicelessly aware of its movement. It is not that the speaker is directing you to look or to listen to all the inward movement of consciousness in a particular way. Consciousness is common to all mankind. Throughout the world man suffers inwardly as well as outwardly, there is anxiety, uncertainty, utter despair of loneliness; there

is insecurity, jealousy, greed, envy and suffering. Human consciousness is one whole; it is not your consciousness or mine. This is logical, sane, rational: wherever you go, in whatever climate you live, whether you are affluent or degradingly poor, whether you believe in god, or in some other entity, belief and faith are common to all mankind—the images and symbols may be totally different in various localities but they stem from something common to all mankind. This is not a mere verbal statement. If you take it as a verbal statement, as an idea, as a concept, then you will not see the deep significance involved in it. The significance is that your consciousness is the consciousness of all humanity because you suffer, you are anxious, you are lonely, insecure, confused, exactly like others, though they live ten thousand miles away. The realization of it, the feeling of it—the feeling in your guts—is totally different from the mere verbal acceptance. When you realize that you are the rest of mankind, it brings a tremendous energy, you have broken through the narrow groove of individuality the narrow circle of me and you, us and them. We are examining together this very complex consciousness of man, not the European man, not the Asiatic man, but this extraordinary movement in time that has been going on in consciousness for millions of years.

转折点就在我们的意识中。我们的意识是一个非常复杂的东西，关于它，东方和西方都曾有过大量的描述。我们对于自己的意识没有觉察。要检查意识的复杂性，你必须自由地看，无选择地觉察它的运作。不是建议你以一种特别的方式去看或听意识的一切内在运动。意识是一切人类共有的。全世界的人都承受着内在和外在的痛苦，包括焦虑、迷茫、对孤独的彻底绝望，以及不安全感、羡慕、贪婪、嫉妒和悲苦。人类的意识是一个整体，它不是你的意识或我的意识。它是有逻辑的、明智的、理性的，无论你去什么地方，无论你生活在哪种气候中，无论你是富有还是贫穷，无论你信仰神明还是某个别的实体，信念和信仰是全人类共有的——不同地区的形象和符号或许完全不同，但是它们都源于某种人类共有的东西。这不只是一个口头上的陈述。如果你把它作为一个口头陈述、一个想法、一个观念，那么你就不会看到其中所包含的深刻意义。这个意义是，你的意识是所有人类的意识，因为你痛苦、忧虑、孤独、没有安全感、困惑，这完全和别人一样，尽管他们生活在万里之外。对它的认识和感受——你内心的感受——完全不同于口头上的赞同。当你认识到你和其他所有的人一样，这会带给你惊人的活力，因为你打破了个体的狭隘常规，打破了我和你、我们和他们的狭隘循环。我们正在共同探寻这一非常复杂的人类意识，不只是欧洲人、亚洲人，而且这一项卓越的活动已经持续了数百万年。

>> Please do not accept what the speaker is saying; It will have no meaning if you do it. If you do not begin to doubt, begin to question, be sceptical to enquire, if you hold on to your own particular belief, faith, experience or the

accumulated knowledge, then you will reduce it all to some kind of pettiness with very little meaning. If you do that you will not be facing the tremendous issue that is facing man.

请不要轻易接受讲话者所说的，那样做将毫无意义。如果你没有开始怀疑，开始质疑，带着怀疑去探究；如果你总是坚持自己的特定信念、信仰、经验或者积累的知识，你就会把这一切简单地看成某种微不足道的东西。这样，你就不会正视人类面临的巨大问题。

>> We have to see what our actual consciousness is. Thought and all the things that thought has put together, is part of our consciousness—the culture in which we live, the aesthetic values, the economic pressures, the national inheritance. If you are a surgeon or a carpenter, if you specialize in a particular profession, that group consciousness is part of your consciousness. If you live in a particular country with its particular tradition and religious culture,

that particular group consciousness has become part of your consciousness. These are facts. If you are a carpenter you have to have certain skills, understand the nature of wood and the tools of the trade, so you gradually belong to a group that has cultivated these special skills and that has its own consciousness—similarly the scientist, the archeologist, just as the animals have their own particular consciousness as a group. If you are a housewife you have your own particular group consciousness, like all the other housewives. Permissiveness has spread throughout the world; it began in the far West and has spread right through the world. That is a group-conscious movement. See the significance of it; go into it for yourself, see what is involved in it.

我们必须看到意识的真实面貌。思想和思想构成的一切是我们意识的一部分——我们赖以生存的文化、审美观念、经济压力、民族传承。如果你是一个外科医生或一个木匠，如果你从事一种特殊的职业，那个群体的意识也是你意识的一部分。如果你生活在一个有特殊的传统和宗教文化的国家，那个特殊的群体意识就成了你意识的一部分。这些都是事实。如果你是一个木匠，你必须拥有某种技能，了解木材的性质和使用的工具，然后你会渐渐地融入一个培养这些技能的团体——类似于科学家和考古学家，这个团体拥有它们自己的意识。作为一个群体，连动物都有它们自己特殊的意识。如果你是一个家庭主妇，你会和所有其他家庭主妇一样，拥有自己特殊的群体意识。性解放蔓延至全世界，它开始于遥远的西方，并向整个世界传播。那是一个群体意识的活动。去看到它的意义，亲自探究它，看到它所包含的内容。

>> Our consciousness includes, in the much deeper consciousness, our fears. Man has lived with fear for generation after generation. He has lived with pleasure, with envy, with all the travail of loneliness, depression and confusion; and with great sorrow, with what he calls love and the everlasting fear of death. All this is his consciousness which is common to all mankind. Realize what it means:

it means that you are no longer an individual. This is very hard to accept because we have been programmed, as is the computer, to think we are individuals. We have been programmed religiously to think that we have souls separate from all the others. Being programmed our brain works in the same pattern century after century.

在更为深入的层面，我们的意识包含恐惧。人类带着恐惧生活了一代又一代。我们始终带着快乐，带着嫉妒，带着孤独、沮丧和困惑导致的所有痛苦，带着巨大的悲伤，带着所谓的爱，以及对死亡的无尽恐惧。这一切都是人类共有的意识。要认识到这意味着什么：这意味着你不再是个体。这很难接受，因为我们一开始也像计算机一样被设定了，我们似乎拥有不同于其他所有人的灵魂。我们的脑子被设定了，它以相同的模式运作了一个又一个世纪。

>> So, how shall a human being—who is actually the rest of mankind—how shall he face this crisis, this turning

point? How will you as a human being, who has evolved through millennia upon millennia, thinking as an individual, a turning point, see what actually is and in that very perception move totally in another direction?

那么，一个人——他事实上也是其他所有的人——将会如何面对这个危机、这个转折点？作为进化了数千年的人类，在这个转折点，你将如何从个体的角度来思考，如何看到世界的本质，如何从全新的角度去觉察？

>> Let us understand together what it means to look—to look at the actuality of thought. You all think, that is why you are here. You all think and thought expresses itself in words, or through a gesture, through a look, through some body movement. Words being common to each one of us, we understand through those words the significance of what is being said. Yet thought is common to all mankind—it is a most extraordinary thing if you have discovered

that, for then you see that thought is not your thought, it is thought. We have to learn how to see things as they actually are—not as you are programmed to look. Can we be free of being programmed and look? If you look as a Christian, a Catholic or a Protestant—which are all so many prejudices—then you will not be able to understand the enormity of the danger, the crisis, that we are facing. If you belong to a certain group, or follow a certain guru, or are committed to a certain form of action, then, because you have been programmed, you will be incapable of looking at things as they actually are. It is only if you do not belong to any organization, to any group, to any particular religion or nationality, that you can really observe. If you have accumulated a great deal of knowledge from books and from experience, your mind has already been filled, your brain is crowded with experience, with your particular tendencies and so on—all that is going to prevent you from looking. Can we be free of all that to look at what is actually happening in the world?—At the terror and the terrible religious sectarian divisions, one guru opposed to another idiotic guru, the vanity behind all that, the power, the position, the wealth of these gurus, it is appalling. Can you look at yourself—not as a separate human being but as a human being who is actually the rest of mankind?

让我们共同了解“看”意味着什么——去看思想的实质。你们都会思考，这是你们在这里的原因。你们都会思考，思考用言辞，或通过手势、眼神、某种身体动作来表达自己。我们在场的人语言是共同的，我们通过语言来理解所说内容的含义。然而思想是所有人共有的，它是最不寻常的事情。如果你发现这点的话，那时你就会看到，思想不是你的，它只是思想。我们必须学会如实地而不是以你被设定的样子去看待事情。我们能够免于被设定地去看吗？如果你仅作为一个基督徒、天主教徒或新教徒——他们有太多的偏见——去看，你就不能了解我们所面临的危险、危机的严重性。如果你属于一个特定的团体，或者跟随特定的古鲁，或投身于特定的行动，那么因为你被设定了，所以你将不能如实地看事情。只有不属于任何组织、任何团体、任何特定的宗教或国家，你才能真正观察。如果你从书本和经验中积累了大量知识，你的心已经被填满了，你的脑子充斥着经验和特殊的倾向等等，这些都将妨碍你去看。我们能否摆脱那些东西，去看世界上实际发生的事情——可怕的教派分裂，一个古鲁反对另一个古鲁，这一切背后的虚浮，这些古鲁的权力、地位、财富，是骇人听闻的。你能审视你自己吗——不是作为个体，而是和其他所有人一样的人？

>> When you are able to see clearly, without any distortion, then you begin to enquire into the nature of consciousness, including the much deeper layers of consciousness. You have to enquire into the whole movement of thought, because it is thought that is responsible for all the content of consciousness, whether it is the deep or the superficial layers. If you had no thought there would be no fear, no sense of pleasure, no time; thought is responsible. All the achievements of the great painters, poets, composers, are

the activity of thought: the composer, inwardly hearing the marvellous sound, commits it onto paper. That is the movement of thought. Thought is responsible for all the gods in the world, all the saviours, all the gurus; for all the obedience and devotion; the whole is the result of thought which seeks gratification and escape from loneliness. Thought is the common factor of all mankind. The poorest villager in India thinks as the chief executive thinks, as the religious leader thinks. That is a common everyday fact. That is the ground on which all human beings stand. You cannot escape from that.

当你能毫不扭曲地、清晰地看，你就会开始探询意识的本质，包括意识的更深层面。你必须探究思想的整个活动，因为正是思想要对意识负责，无论是在深刻的还是在肤浅的层面。如果你没有思想，就不会有恐惧，不会有快乐的感觉，不会有时间。思想是负有责任的。伟大的画家、诗人、作曲家的一切成就都是思想的活动：作曲家在内心听到不可思议的声音，在纸上记录下来，这就是思想的活动。思想要对世界上存在的所有的神、所有的救世主、所有的古鲁负责，因为所有的服从和奉献全都是思想寻求满足和逃避孤独的结果。思想是人类的共同要素。印度最贫困的村民与国家元首及宗教领袖一样思考。这是一个普通的常识，是人类的共同立场。你无法脱离它。

>> Thought has done marvellous things to help man but it has also brought about great destruction and terror in the world. We have to understand the nature and the movement of thought; why you think in a certain way; why you cling to certain forms of thought; why you hold on to certain experiences; why thought has never understood the nature of death. We have to examine the very structure of thought—not your thought because it is fairly obvious what your thought is, for you have been programmed. But if you enquire seriously into what thinking is, then you enter into quite a different dimension—not the dimension

of your own particular little problem. You must understand the tremendous movement of thought, the nature of thinking—not as a philosopher, not as a religious man, not as a member of a particular profession, or a housewife—the enormous vitality of thinking.

虽然思想帮助人类做过不可思议的事情，但也为世界带来了巨大的破坏和恐惧。我们必须理解思想的本质和运行方式，你为什么以特定的方式思考，为什么抓住特定的思考形式不放，为什么坚持特定的经验，为什么思想从来无法理解死亡的本质。我们必须检查思想的结构——不是你的思想，因为你的思想是什么样子是相当明显的，因为你是被设定的。如果你认真探究思想是什么，你就会进入一个相当不同的维度，不是你自己特殊这个小问题的维度。你必须理解思想的惊人活动、思想的本质、思想的巨大活力——不是作为一个哲学家、一个虔诚的宗教信徒、一个特殊职业中的一员或者一个家庭主妇拥有的思想。

>> Thought is responsible for all the cruelty, the wars, the war machines and the brutality of war, the killing, the terror, the throwing of bombs, the taking of hostages in the name of a cause, or without a cause. Thought is also responsible for the cathedrals, the beauty of their structure, the lovely poems; it is also responsible for all the technological development, the computer with its extraordinary capacity to learn and go beyond man's thought. What is thinking? It is a response, a reaction, of memory. If you had no memory you would not be able to think. Memory is stored in the brain as knowledge, the result of experience. This is how our brain operates. First, experience; that experience may have been from the beginning of man, which we have inherited, that experience gives knowledge which is stored up in the brain; from knowledge there is memory and from that memory thought. From thought you act. From that action you learn more. So you repeat the cycle. Experience, knowledge, memory, thought, action; from that action learn more and repeat. This is how we are

programmed. We are always doing this: having remembered pain, in the future avoid pain by not doing the thing that will cause pain, which becomes knowledge, and repeat that. Sexual pleasure, repeat that. This is the movement of thought. Thought says to itself: "I am free to operate." Yet thought is never free because it is based on knowledge and knowledge is obviously always limited. Knowledge must also be always limited because it is part of time. I will learn more and to learn more I must have time. I do not know Russian but I will learn it. It may take me six months or a year or a lifetime. Knowledge is the movement of time. Time, knowledge, thought and action; in this cycle we live. Thought is limited, so whatever action thought generates must be limited and such limitation must create conflict, must be divisive.

思想要对所有残忍、战争、战争机器和战争中的残暴、杀戮、恐怖、投掷炸弹、为了任何原因或没有任何原因地扣留人质负责。思想也要对大教堂的完美构造和迷人诗句负责，还要对所有的技术进步负责，对计算机惊人的学习能力和超过人类的思考能力负责。思考是什么？它是一种记忆的反应、回应。如果没有记忆，你就不能思考。作为知识和经验的结果，记忆被储存在脑子里。这就是我们脑子运作的方式。那个经验或许来自人类的早期，被我们继承下来，那个经验提供了储存在脑子里的知识；记忆来自知识，而思想来自记忆，思想产生了行动，从这个行动中你学到更多。于是你重复着这个循环。经验、知识、记忆、思想、行动；由行动学会更多，然后重复。我们就是这样被设定的。我们总是这样：记住痛苦，通过在将来不去做会导致痛苦的事情来避免痛苦，这变成了知识，然后重复它。性是快乐的，重复它。这就是思想的活动。思想对自己说“我在自由运作”，然而思想从未自由，因为它是基于知识的，而知识明显是受限的。所以思想也一定是受限的，因为它是时间的一部分。我将学会更多，而要学会更多我就必须有时间。我不懂俄语，但是我要学会它，那或许需要花费我六个月或一年乃至一生的时间。知识是时间的运动。时间、知识、思想和行动，我们活在这个循环当中。思想是受限的，因此思想产生的任何行动必定都是受限的，而这样的限制一定会制造冲突，一定会造成分裂。

>> If I say that I am a Hindu, that I am Indian, I am limited and that limitation brings about not only corruption but conflict because another says, "I am a Christian" or "I am a Buddhist", so there is a conflict between us. Our life from birth to death is a series of struggles and conflicts from which we are always trying to escape, which again causes more conflicts. We live and die in this perpetual and endless conflict. We never seek out the root of that conflict, which is thought, because thought is limited. Please do not ask, "How am I going to stop thought?"—that is not the point. The point is to understand the nature of thought, to look at it.

12th July, 1981

031

如果我说我是印度教徒、是印度人，那么我就是受限制的，而且这种限制不仅造成了腐败，还导致了冲突，因为如果另一个人说“我是基督徒”或“我是佛教徒”，那我和他之间就会有冲突。从生到死，我们的生活是一系列的挣扎和冲突，对此我们总是试图逃避，而这又导致了更多的冲突。我们在这种无尽的冲突中生活并死去。我们从未找到那些冲突的根源，其实根源就是思想，因为思想是受限的。请别问“我该如何阻止思想的产生”，那不是重点。重点是理解思想的本质，去看它。

一九八一年七月十二日

>> We were saying that human consciousness is similar in all human beings. Our consciousness, whether we live in the East or West, is made up of many layers of fears, anxieties, pleasures, sorrows and every form of faith. Occasionally, perhaps, in that consciousness there is also love, compassion, and from that compassion a totally different kind of

intelligence. And always there is the fear of ending, death. Human beings throughout the world from time immemorial have tried to find out if there is something sacred, beyond all thought, something incorruptible and timeless.

我们说到，人类的意识都是相似的。无论我们生活在东方还是西方，我们的意识都是由很多层面的恐惧、焦虑、快乐、悲伤和各种形式的信仰构成的。在那个意识当中，或许会有爱、慈悲，从那份慈悲当中或许会产生一种完全不同的智慧，当然也总是存在对于结束和死亡的恐惧。有史以来，全世界的人都在试图发现某种神圣的、超越思想的东西，某种不会衰朽的、永恒的东西。

>> There are various group consciousnesses; the businessmen with their consciousness, the scientists with theirs and the carpenters with theirs, these are of the content of consciousness and are the product of thought. Thought has created wonderful things; from the extraordinary technology of computers, to telecommunication, to robots, surgery and medicine. Thought

has invented religions; all the religious organizations throughout the world are put together by thought.

有各种各样的团体意识。商人有商人的意识，科学家有科学家的意识，木匠也有自己的意识，这些都是意识的内容，是思想的产物。思想制造了惊人的东西，从非凡的计算机技术，到电信技术，到机器人，再到外科手术和医药。思想催生了宗教，全世界所有的宗教组织都是由思想连接在一起的。

>> Thought has invented the computer. You must understand the complexity and the future of the computer; it is going to outstrip man in his thought; it is going to change the structure of society and the structure of government. This is not some fantastic conclusion of the speaker, or some fantasy, it is something that is actually going on now, of which you may not be aware. The computer has a mechanical intelligence; it can learn and invent. The computer is going to make human labour practically

unnecessary—perhaps two hours work a day. These are all changes that are coming. You may not like it, you may revolt against it, but it is coming.

思想使计算机得以发明。你必须理解计算机的复杂，它将在思考能力方面超过人类，同时改变社会结构和政府结构。这不是讲话者的一些不切实际的论断或某种幻想，它是现在正实际发生的事情，对此你或许没有意识到。计算机拥有机械的智力，它能够学习和发明。计算机将会使人类劳动变得几乎没有必要——也许每天只需要工作两小时。这些改变都正在到来。你可能不喜欢它，你可能想反抗它，但是它正在到来。

>> Thought has invented the computer, but human thought is limited and the mechanical intelligence of the computer is going beyond that of man. It is going to totally revolutionize our lives. So what will a human being be then? These are facts, not some specialized conclusions of the speaker.

思想使计算机得以发明，但是人类的思想是有限的，而计算机的机械智力将超越人类，它将彻底改变我们的生活。那个时候，人将会是什么呢？这些都是事实，不是讲话者的一些专业化的结论。

>> When we consider what the capacity of the computer is, then we have to ask ourselves: what is a human being to do? The computer is going to take over most of the activities of the brain. And what happens to the brain then? When a human being's occupation is taken over by the computer, by the robot, what becomes of the human? We human beings have been "programmed" biologically, intellectually, emotionally, psychologically, through millions of years, and we repeat the pattern of the programme over and over again. We have stopped learning: and we must enquire if the human brain, which has been programmed for so many centuries, is capable of learning and immediately transforming itself into a totally different dimension. If we are not capable of that,

the computer, which is much more capable, rapid and accurate, is going to take over the activities of the brain. This is not something casual, this is a very very serious, desperately serious matter. And we, if we are not aware of what is happening, we will follow that new structure which has been turned out by the computer. See the seriousness of all this, please.

当我们考虑计算机的能力时，我们不得不问自己：一个人该做什么？计算机将会接管大部分的大脑活动，然后大脑会怎么样呢？当一个人的工作被计算机、机器人接管，人会成为什么？数百万年来，在生理上、心智上、情感上、心理上，我们人类始终在被设定，并且我们一再地重复这个程序模式。我们一度停止了学习，但我们现在必须探究，被设定了那么多个世纪的人类头脑，是否有能力学习，并立即转变进入完全不同的维度。如果我们没有那样的能力，那么更加有能力的、快速的和精确的计算机将接管头脑的活动。这不是某种偶然，这是一个极为严肃的事件。而如果我们没有意识到正在发生什么，我们将遵循那个由计算机制造的新的格局。请看到这一切的严重性。

>> Our consciousness has been programmed for thousands and thousands of years to think of ourselves as individuals, as separate entities struggling, in conflict from the moment we are born until we die. We are programmed to that. We have accepted that. We have never challenged it; we have never asked if it is possible to live a life absolutely without conflict. Never having asked it we will never learn about it. We repeat. It is an innate part of our existence to be in conflict—nature is in conflict: that is our argument—and we consider that progress is only through conflict. We are questioning very seriously whether there is an individual consciousness; whether you, as a human being, have a separate consciousness from the rest of mankind. You have to answer this, not just play with it.

我们的意识数千年来都被设定为，我们自己是个体、是单独的实体，从出生的那一刻一直到死，都在挣扎，都处于冲突之中。我们被设定成那样。我们接受了它。我们从未质疑它，我们从未问过自己有没有可能过一种完全没有冲突的生活。因为从未问过，所以我们永远也不会了解它。我们不断重复。冲突是我们生活的固有部分——自然就是冲突的，那是我们的依据——我们认为进步只能通过冲突实现。我们在非常严肃地质疑是否真的存在个体意识；作为一个人，你是否拥有独立于其他人的意识？你必须回答这个问题，不要只是把它当消遣。

>> Having been brought up, programmed, conditioned, to be individuals, then our consciousness is all this activity of thought. Fear and the pursuit of pleasure are the movement of thought. The suffering, anxiety, uncertainty and the deep regret, wound, the burden of centuries of sorrow, are all part of thought. Thought is responsible for what we call love, which has become sensual pleasure, something to be desired.

被培养、设定、制约、个体化，我们的意识就是所有这些思想的活动。恐惧和对快乐的追求也是思想的活动。痛苦、焦虑、不确定以及深刻的惆怅、创伤、背负的多个世纪的悲伤，都是思想的组成部分。思想促成了我们所谓的爱，它已经变成了肉体上的快感、某种被渴求的东西。

>> As we said, and we will repeat it over and over again until we are quite sure of it, we are thinking together, the speaker is not telling you what to think. He is not making propaganda—it is a horrible thing, propaganda. He is not telling you how to act, what to believe, but together, we are investigating the catastrophe that is taking place in the world outside of us, the utter ruthlessness and violence, and also inwardly in each human being the extraordinary conflict that is going on. Together we are examining. It is not—if one may point out—that you are merely listening to some ideas or conclusions; we are not talking about ideas, conclusions or beliefs. We are

looking at this world that human beings have produced, for which all of us are responsible. We must be clear in our understanding—at whatever level that understanding be, whether it is intellectual understanding, which is merely verbal, or the understanding of deep significance so that that understanding acts—that we have come to a point where we have to make a decision, not by the exercise of will, but the decision that will naturally come when we begin to understand the whole nature and structure of the world, both externally and internally. That perception will bring about a decision, an action.

我们说过，而且我们将重复直到我们非常确定，我们是在共同思考，而不是讲话者在告诉你们应该去思考什么。他不是在做宣传——那是可怕的事情。他不是在告诉你们怎么行动、相信什么，而是我们在一起审视周围的世界正在发生的灾难、彻底的冷酷和暴力，以及每个人内心正在发生的惊人的冲突。我们是在一同检查。我们检查的不是你听到的一些想法或结论——

如果可以指出的话——我们不是在谈论想法、结论或信仰。我们是在看人类创造的这个世界，我们所有人都对它负有责任。我们必须清楚地了解——这了解包括任何水平上的，无论是智力上还是言辞上，或是深刻意义上，也正是这种了解产生了行动——我们已经来到了一个必须做决定的转折点，这个决定并非意志强加于你的。而是当你开始从内在和外在了解世界的本质和结构，那个决定会自然地到来。这份觉察会带来一个决定、一个行动。

>> Thought has created the problems which surround us and our brains are trained, educated, conditioned, to the solving of problems. Thought has created the problems, like the division between nationalities. Thought has created the division and the conflict between various economic structures; thought has created the various religions and the divisions between them and therefore there are the conflicts. The brain is trained to attempt to solve these conflicts which thought has created. It is essential that we understand deeply

the nature of our thinking and the nature of our reactions which arise from our thinking. Thought dominates our lives, whatever we do; whatever action takes place, thought is behind that action. In every activity, whether it is sensual or intellectual, or biological, thought is operating all the time. Biologically, through centuries, the brain has been programmed, conditioned—the body acts in its own way, the action of breathing, the beat of the heart and so on.

思想制造了我们身边的问题，为了解决问题，我们的脑子受到训练、教育和约束。思想制造了问题，例如国家之间的隔阂。思想制造了各种经济结构之间的分隔和冲突，思想制造了各种宗教之间的分隔，因而也制造了冲突。而我们的脑子受到训练，试图去解决这些思想制造的冲突。深入理解我们思维的本质，以及从我们的思维中产生的反应的本质，对我们来说是必要的。思想占据着我们的生活，无论我们做什么，无论发生什么，其背后都是思想。在每一个活动中，无论是感性的、理性的还是生物性的，思想都一直在起作用。在生物学上，许多世纪以来，大脑被设定、约束——身体以它自己的方式运动，如呼吸、心跳等等。

>> Thought is a movement in time and space. Thought is memory, the remembrance of past things. Thought is the activity of knowledge, knowledge which has been gathered together through millions of years and stored as memory in the brain. If you observe the activity of your thinking, you will see that experience and knowledge are the basis of your life. Knowledge is never complete, it must always go together with ignorance. We think knowledge is going to solve all our problems, whether the knowledge of the priest, the guru, the scientist, the philosopher, or the psychiatrist. But we have never questioned whether knowledge in itself can solve any of our problems—except perhaps technological problems.

思想是时间和空间中的运动。思想是记忆，对过去的事情的回忆。思想是知识的活动，是经过数百万年汇集在一起、以记忆的形式储存在大脑里的知识。如果你观察你的思想活动，你就会看到，经验和知识是你生活的基础。知识从来不是完整的，它必然总是与无知同行。我们认为知识将会解决我们所有的问题，无论是牧师、古鲁、科学家、哲学家的知识还是精神科医生的知识。但是我们从未怀疑知识本身能否解决我们的任何问题——或许除了技术问题。

>> Knowledge comes through time. To learn a language you need time. To learn a skill or to drive a car efficiently takes time. The same movement of time is brought over to the psychological field; there too we say, "I must have time to learn about myself." "I must have time in order to change myself from 'what I am' to 'what I should be'. " Bringing over the activity of the external world into the psychological world means that time is a great factor in our life—tomorrow, the past and the present. Time is thought.

Time is required in the acquisition of knowledge through experience, both externally in the world and inwardly. That is the way we have been programmed.

知识是经由时间而来的，学会一种语言需要时间，学会一门技术或高效地驾驶一辆汽车也需要时间。关于“时间”的运动也波及了心理学领域，在那个领域我们也说“我必须花时间了解自己”“我必须花时间把自己从‘现在如何’变成‘应该如何’”。将外在世界的活动带进精神世界意味着时间是我们生活中的重要元素——未来、过去和现在。时间即是思想。在凭借经验获取知识方面，时间是必需的，无论外在还是内在。我们就是那样被设定的。

>> Being so programmed we consider time is necessary to bring about a deep, fundamental change in the human structure. We employ time as thought—“I am this, I shall be that.” You would also say in the technical world: “I do not know how to construct a computer but I will learn.”

Time, knowledge, memory, thought, they are a single unit; they are not separate activities but a single movement. Thought, the outcome of knowledge, must everlastingly be incomplete and therefore limited, because knowledge is incomplete. Whatever is limited must bring about conflict. Nationality is limited. Religious belief is limited. An experience which you have had, or which you are longing for, is limited. Every experience must be limited.

因为受到这样的设定，我们认为想要深刻地、根本性地转变人性是需要时间的。我们花费时间来思考——“我是这个，我将成为那个”。在技术世界你也会说：“我不知道如何装配计算机，但我会学习。”时间、知识、记忆、思想是一体的，它们不是分开的活动，而是同一个活动。思想作为知识的产物，一定永远是不完整的，因而是有限的，因为知识是不完整的。任何有限的东西都会带来冲突。国家是有限的，宗教信仰是有限的，你有过的经验或者你渴望的经验是有限的。每一个经验必定都是有限的。

>> **Questioner:** Why?

Krishnamurti: Because there are more experiences. I may have an experience sexually, or the experience of the possession of wealth, the experience of giving everything up and going into a monastery—those experiences are all limited.

Thought, being limited, creates problems—national, economic and religious divisions; then thought says, "I must solve them." So thought is always functioning in the resolution of problems. And the computer, a mechanism which has been programmed, can outstrip all of us because it has no problems; it evolves, learns, moves.

提问者：为什么？

克里希那穆提：因为存在更多的经验。我也许有性的经验，有获得财富的经验，有放弃一切进入修道院的经验——但那些经验都是有限的。

思想，由于受限，制造了问题——民族的、经济的和宗教的分隔。然后思想说“我必须解决它们”，因此思想总是在解决问题中运作。而计算机，这种被设定程序的机械装置，能够胜过我们所有人，因为它没有问题，并且不断优化、学习、运行。

>> Our consciousness has been programmed as an individual consciousness. We are questioning whether that consciousness, which we have accepted as individual, is actually individual at all. Do not say: "What will happen if I am not an individual?" Something totally different may happen. You may have an individual training in a particular trade, in a particular profession, you may be a surgeon, an engineer, but that does not make you an individual. You may have a different name, a different form—that does not make individuality; nor the acceptance that the brain through time has affirmed: "I am an individual, it is my desire to fulfil, to become through struggle··· " That so-called individual consciousness, which is yours, is the consciousness of all humanity.

我们的意识被设定为个体的意识。我们质疑它实际上到底是不是个体的。不要说“如果我不是个体，那么会发生什么”，或许会发生某种完全不同的事情。你也许在特别的行业或职业中受过个人培训，你或许是一位外科医生、一位工程师，但是那并没有使你成为个体。你或许有一个不同的名字、一个不同的形式，但那不会带来个体性。接受脑子通过时间确定“我是个体，我渴望通过努力去完成、去成为……”这点也不会使你成为个体，那个所谓的个体意识，它是你的，但也是全人类的意识。

>> If your consciousness, which you have accepted as separate, is not separate, then what is the nature of your consciousness? Part of it is the sensory responses. Those sensory responses are naturally, necessarily, programmed to defend yourself, through hunger to seek food, unconsciously. Biologically you are programmed. Then the content of your consciousness includes many hurts

and wounds that you have received from childhood, many forms of guilt; it includes the various ideas, imaginary certainties; many experiences, both sensory and psychological; there is always the basis, the root, of fear in its many forms. With fear naturally goes hatred. Where there is fear there must be violence, aggression, the tremendous urge to succeed, both in the physical and the psychological world. In the content of consciousness there is the constant pursuit of pleasure; the pleasure of possession, of domination, the pleasure of a philosopher with his immense knowledge, the guru with his circus. Pleasure again has innumerable forms. There is also pain, anxiety, the deep sense of abiding loneliness and sorrow, not only the so-called personal sorrow but also the enormous sorrow brought about through wars, through neglect, through this endless conquering of one group of people by another. In that consciousness there is the racial and group content; ultimately there is death.

如果你认为自己拥有独立意识，实际它并不是独立的，那么它的本质是什么呢？一部分是感官反应。那些感官反应是自然的、必要的、被设定来保护你的，由于饥饿去觅食，这是无意识的。这在生物学上是被设定的。你的意识包含你从童年开始经受的许多痛苦和创伤、自责、各种想法、想象的现实、感受和心理上的经验，以及导致各种恐惧的基础和根源。恐惧自然会导致仇恨。恐惧一定会在现实世界和心理世界带来暴力、攻击和巨大的成功欲。意识中存在着对快乐的无尽追求，占有的快乐、控制的快乐、知识渊博的哲学家的快乐、会变戏法的古鲁的快乐。快乐有数不清的形式。当然意识中还有痛苦、焦虑、深刻而持久的孤独和伤悲。不只是有所谓个人的伤悲，还有战争、疏忽、一个群体被另一个群体无尽地征服带来的巨大的伤悲。意识中还存在种族和群体性，最后还有死亡。

>> This is our consciousness—beliefs, certainties and uncertainties, anxieties, loneliness and endless misery. These are the facts. And we say this consciousness is mine! Is that so? Go to Asia, America, Europe, anywhere where human beings are; they suffer, they are anxious, lonely, depressed, melancholic, struggling and in conflict—they are just the same as you. So, is your consciousness different from that of another? I know it is very difficult for people to accept—you may logically accept it, intellectually you may say, "Yes, that is so, maybe". But to feel this total human sense that you are the rest of mankind requires a great deal of sensitivity. It is not a problem to be solved. It is not that you must accept that you are not an individual, that you must endeavour to feel this global human entity. If you do, you have made it into a problem which the brain is only too ready to try to solve! But if you really look at it with your mind, your heart, your whole being totally aware of this fact, then you have broken the programme. It is naturally broken. But if you say, "I will break it", then you are again

back into the same pattern. To the speaker this is utter reality, not something verbally accepted because it is pleasant; it is something that is actual. You may have logically, reasonably and sanely examined and found that it is so; but the brain which has been programmed to the sense of individuality is going to revolt against it. The brain is unwilling to learn. Whereas the computer will learn because it has nothing to lose. But here you are frightened of losing something.

这就是我们的意识——信念、确定性和不确定性、忧虑、孤独和无尽的哀伤。这些都是事实。而我们却说这个意识是我的！是这样吗？亚洲、美洲、欧洲，任何地方的人都是如此。他们痛苦、忧虑、孤独、沮丧、抑郁、挣扎、充满冲突——和你一样。那么，你的意识和另一个人的意识不同吗？我知道，对人们来说，这很难接受——或许逻辑上你接受它，但理智上你仍会怀疑，“是的，是那样的，也许吧”。但是要感受到人类是一体的，要感受到自己就是其他的人，这需要敏锐的感觉。它不是一个要被解决的问题，不是说你必须接受你不是个体，必须去尽力感受这个全球性的人类实体。因为如果你这样做，你就使它成了一个头脑想要尝试去解决的问题！但是如果你真的用你的智慧、你的心、你的整个意识完全地去觉察这个事实，那么你就打破了那个程序设定。它自然就被打破了。但如果你说“我要打破它”，那么你又回到了相同的模式。对讲话者来说这绝对是事实，而不是因令人愉快而容易在口头上接受的东西；它是实际的东西。你也许会从逻辑上、道理上明智地检查它并发现它是如此，但是被设定为个体意识的脑子会反抗它。脑子不愿意学习，而计算机愿意学习，因为它不会失去什么。但是你们都害怕失去。

>> Can the brain learn? That is the whole point; so now we have to go into this question of what learning is. Learning for most of us is a process of acquiring knowledge. I do not know the Russian language but I will learn it. I will learn day after day, memorizing, holding on to certain words, phrases and the meanings, syntax and grammar. If

I apply myself I can learn almost any language within a certain time. To us, learning is essentially the accumulation of knowledge or skill. Our brains are conditioned to this pattern. Accumulate knowledge and from that act. When I learn a language, there knowledge is necessary. But if I am learning psychologically about the content of my mind, of my consciousness, does learning there imply examining each layer of it and accumulating knowledge about it and from that knowledge acting—following the same pattern as learning a language? If the brain repeats that pattern when I am learning about the content of my consciousness, it means that I need time to accumulate knowledge about myself, my consciousness. Then I determine what the problems are and the brain is ready to solve them—It is repeating this endless pattern and that is what I call learning. Is there a learning which is not this? Is there a different action of learning, which is not the accumulation of knowledge? You understand the difference?

头脑能学习吗？这是重点。所以现在我们必须讨论“什么是学习”这个问题。对我们大多数人来说，学习是一个获取知识的过程。我不懂俄语，但我将学习它。我将日复一日地学习，熟记、掌握特定的单词、短语，以及词语含义、结构和语法。如果我专心致志，我就能在一定时间内学会任何一门语言。对我们来说，学习本质上就是知识或技能的积累。我们的大脑被设定为这样的模式，积累知识并展开行动。当我们学习一门语言的时候，知识是必要的。但是如果我们学习有关心灵的、意识的内容，那它意味着检查相关知识的每一个层面、收集有关的知识并且按照那个知识行动，就和学习语言一样吗？当我了解我的意识时，如果大脑重复那个模式，这意味着我需要时间去收集关于我和我的意识的知识。然后我确定问题是什么，大脑去解决它们。它重复着这个无尽的模式，这是我所谓的学习。有没有一种学习不是这样的？有没有一种不同的学习行为，它不是知识的积累。你了解这个不同吗？

>> Let me put it differently: from experience we acquire knowledge, from knowledge memory; the response of memory is thought, then from thought action, from that action you learn more, so the cycle is repeated. That is the pattern of our life. That form of learning will never solve our problems because it is a repetition. We acquire more knowledge which may lead to the better action; but that action is limited and this we keep repeating. The activity from that knowledge will not solve our human problems at all. We have not solved them, it is so obvious. Over millions of years we have not solved our problems: we are cutting each other's throats, we are competing with each other, we hate each other, we want to be successful, the whole pattern is repeated from the time man began and we are still at it. Do what you will along this pattern and no human problem will be solved, whether it be political, religious or economic, because it is thought that is operating.

让我换个说法：我们从经验中获得知识，从知识中产生记忆，记忆对应的反应是思想，然后思想产生行动，从那个行动中你学到更多，这个循环是重复的。这就是我们生活的模式。这种学习的形式永远也不会解决我们的问题，因为它是重复的。我们获取更多的知识，那或许会带来更好的行动，但是那个行动是受限的，我们还在保持重复。从那个知识中产生的活动根本不会解决我们人类的问题。我们还是没有解决它们，这很明显。数百万年来，我们还是没有解决我们的问题，我们互相残杀，我们互相竞争，我们互相仇恨，我们想要成功，从有人类开始整个模式就在被重复，现在依然如此。沿着这个模式去做，人类的问题不会被解决，无论是政治的、宗教的还是经济的问题，因为它是由思想在控制。

>> Now, is there another form of learning; learning, not in the context of knowledge, but a different form, a non-accumulative perception-action? To find out we have to enquire whether it is possible to observe the content of

our consciousness and to observe the world without a single prejudice. Is that possible? Do not say it is not possible, just ask the question. See whether, when you have a prejudice, you can observe clearly. You cannot, obviously. If you have a certain conclusion, a certain set of beliefs, concepts, ideals, and you want to see clearly what the world is, all those conclusions, ideals, prejudices and so on will actually prevent it. It is not a question of how to get rid of your prejudices but of seeing clearly, intelligently, that any form of prejudice, however noble or ignoble will actually prevent perception. When you see that, prejudices go. What is important is not the prejudice but the demand to see clearly.

那么，是否存在另一种形式的学习，不是在知识的背景中，而是一种不同的形式、一种非积累的觉察活动？要弄清楚这件事，我们就必须探究，有没有可能不携带个人的成见去观察我们的意识、观察这个世界。那可能吗？不要说那不可能，要先考虑这个问题，看一看当你带着成见的时候能否清晰地去观察。显然你不能，如果你有一个确定的结论，一套确定的信仰、观念、理想，而你想清楚地看到这个世界的样子，那你所有的结论、理想、成见等确实会阻碍它。这不是如何去除成见的问题，而是如何清晰地、机智地看的问题，任何形式的成见，无论高尚的还是卑鄙的，都会在事实上妨碍觉察。而当你看到那些，成见就消失了。重要的不是成见，而是需要清晰地看。

>> If I want to be a good surgeon I cannot do so with ideals or prejudices about surgeons; I must actually perform surgery. Can you see that a new form of action, a new form of non-accumulative knowledge, is possible which will break the pattern, break the programme, so that you are acting totally differently?

如果我想成为一名好的外科医生，我就不能对外科医生抱有成见，我必须实际地去完成手术。你能看到那种不需要积累知识的新的行动吗？它可能会打破模式，打破程序，让你以全然不同的方式行动！

>> The way we have lived, over millions of years, has been the repetition of the same process of acquiring knowledge and acting from that knowledge. That knowledge and action is limited. That limitation creates problems and the brain has become accustomed to solving the problems which knowledge has repeatedly created. The brain is caught in that pattern and we are saying that that pattern will never, in any circumstance, solve our human problems. Obviously we have not solved them up till now. There must be a different, a totally different, movement, which is a non-accumulative perception-action. To have non-accumulative perception is to have no prejudice. It is to have absolutely no ideals, no concepts, no faith—because all those have destroyed man, they have not solved his problems.

数百万年以来，我们的生活都重复着相同的过程，就是积累知识并根据知识去行动。知识和行动是局限的，局限造成了问题，而大脑变得习惯于解决这些知识不断制造的问题。脑子陷在了那个模式中，我们说的就是，在任何情况下，那种模式永远都不会解决我们人类的问题。显然，到目前为止我们并没有解决它们。一定有一种不同的、一种完全不同的活动，它是一种非积累的觉察活动。非积累的觉察就是没有成见，就是完全没有理想、没有观念、没有信仰——因为这些都伤害了人类，却没有解决人类的问题。

>> So, have you a prejudice? Have you a prejudice which has something in common with an ideal? Of course. Ideals are to be accomplished in the future, and knowledge becomes tremendously important in the realizing of ideals. So, can you observe without accumulation, without the destructive nature of prejudices, ideals, faiths, beliefs and your own conclusions and experiences? There is group consciousness, national consciousness, linguistic consciousness, professional consciousness, racial consciousness, and there is fear, anxiety, sorrow, loneliness, the pursuit of pleasure, love and finally death. If you keep acting in that circle, you maintain

the human consciousness of the world. Just see the truth of this. You are part of that consciousness and you sustain it by saying, "I am an individual. My prejudices are important. My ideals are essential"—repeating the same thing over and over again. Now the maintenance, the sustenance and the nourishment, of that consciousness takes place when you are repeating that pattern. But when you break away from that consciousness, you are introducing a totally new factor in the whole of that consciousness.

那么，你有成见吗？你有和理想存在共同之处的成见吗？当然有。理想是要在将来完成的，在实现理想的过程中，知识变得极其重要。因此，没有积累，没有具备破坏性本质的成见、理想、信仰、信念，以及自己的结论和经验，你能否去观察？群体意识、国家意识、语言意识、职业意识、种族意识，以及恐惧、焦虑、悲伤、孤独、对快乐的追逐、爱和最终的死亡，如果你始终在那个循环中行动，你就永远拥有全世界的人类意识。看看这个事实吧。你是那个意识的一部分，并且通过说“我是个体，我的成见很重要，我的理想不可或缺”，不断地重复相同的东西使那个意识得以维持。当你重复那个模式的时候，对那个意识的维护、支持和滋养就产生了。但是当你脱离了那个意识，你就往整个意识中带进了一个全新的元素。

>> Now, if we understand the nature of our own consciousness, if we see how it is operating in this endless cycle of knowledge, action and division—a consciousness which has been sustained for millennia—if we see the truth that all this is a form of prejudice and break away from it, we introduce a new factor into the old. It means that you, as a human being who is of the consciousness of the rest of mankind, can move away from the old pattern of obedience and acceptance. That is the real turning point in your life. Man cannot go on repeating the old pattern, it has lost its meaning—in the psychological world it has totally lost its meaning. If you fulfil yourself, who cares? If you become a saint, what does it matter? Whereas, if you totally move away from that

you affect the whole consciousness of mankind.

14th July, 1981

那么，如果我们理解了自身意识的本质，如果我们看到了它如何在这个无尽的知识、行动和分隔的循环中运作——这个被维持了上千年的意识——如果我们看到了这个事实，看到所有这一切都是成见，并从中脱离，我们就将一个新的元素带进来了。这意味着，作为一个拥有其他人意识的人，能够从服从和接受的旧模式中脱离出来，那是你生命中真正的转折。人不能再继续重复旧的模式，它已经失去了意义——在心理世界尤其如此。如果你要满足自己，谁会在乎呢？如果你变成了一个圣人，这有什么关系呢？然而，如果你能完全从那些东西中脱离出来，你就影响了整个人类的意识。

一九八一年七月十四日

第二章

The Whole Universe is an Orderly Existence

整个宇宙是有秩序的

>> I would like to repeat that we are not trying to convince you of anything—that must be clearly understood. We are not trying to persuade you to accept a particular point of view. We are not trying to impress you about anything; nor are we doing any propaganda. We are not talking about personalities, or who is right and who is wrong, but rather trying to think out, to observe, together, what the world is and what we are, what we have made of the world and what we have made of ourselves. We are trying together to examine both the inward and the outward man.

我想再重复一次，我们并不是试图让你相信任何事情——这一点必须说清楚。我们不是要说服你接受一个特别的观点，不是要让你对什么东西产生深刻印象，也不是在做任何宣传。我们不是在谈论人格，或者谁是对的谁是错的，而是想要一起去了解、去观察世界是什么，我们是什么，我们把世界变成了什么，我们把自己变成了什么。我们是要共同去检查内在和外在的人。

>> To observe clearly one must be free to look—obviously. If one clings to one's particular experiences, judgements and prejudices, then it is not possible to think clearly. The world crisis which is right in front of us demands, urges, that we think together so that we can solve the human problem together, not according to any particular person, philosopher, or particular guru. We are trying to observe together. It is important to bear in mind all the time that the speaker is merely pointing out something which we are examining together. It is not something one-sided but rather that we are cooperating in examining, in taking a journey together and so acting together.

要想清楚地观察，你必须自由地看——这很明显。如果你坚持自己特定的经验、看法和成见，那么就不可能清晰地思考。世界性的危机就在我们面前，它要求、催促我们共同思考，从而让我们能够共同解决人类的问题，而不是依靠任何特定的人、哲学家，或特定的古鲁。我们试图共同观察。要时刻记得，讲话者只是在指出一些我们正在一同检视的东西。它不是单方面的行为，而是要我们一起研究、一起踏上旅程，并一起行动。

>> It is very important to understand that our consciousness is not our individual consciousness. Our consciousness is not only that of the specialized group, nationality and so on, but it is also all the human travail, conflict, misery, confusion and sorrow. We are examining together that human consciousness, which is our consciousness, not yours or mine, but ours.

我们的意识不是我们的个体意识，了解这一点非常重要。我们的意识不属于特定群体、特定国家，它是所有人的艰辛、冲突、不幸、困惑和忧伤。我们正在共同检查人类的意识，也就是我们的意识，不是你的或我的，而是我们的。

>> One of the factors that is demanded in this examination is the capacity of intelligence. Intelligence is the capacity to discern, to understand, to distinguish; it is also the capacity to observe, to put together all that we have gathered and to act from that. That gathering, that discernment, that observation, can be prejudiced; and intelligence is denied when there is prejudice. If you follow another, intelligence is denied; the following of another, however noble, denies your own perception, denies your own observation—you are merely following somebody who will tell you what to do, what to think. If you do that, then intelligence does not exist; because in that there is no observation and therefore no intelligence. Intelligence demands doubting, questioning, not being impressed by others, by their enthusiasm, by their energy. Intelligence demands that there be impersonal

observation. Intelligence is not only the capacity to understand that which is rationally, verbally explained but also implies that we gather as much information as possible, yet knowing that that information can never be complete, about anybody or anything. Where there is intelligence there is hesitation, observation and the clarity of rational impersonal thinking. The comprehension of the whole of man, of all his complexities, all his physical responses, his emotional reactions, his intellectual capacities, his affections and his travails, the perceiving of all that at one glance, in one act, is supreme intelligence. Intelligence has not, so far, been able to transcend conflict. We are going together to see if it is possible for the brain to be free from conflict. We live with conflict from the time we are born and will continue to do so until we die. There is the constant struggle to be, to become something spiritually, so–called, or psychologically; to become successful in the world; to fulfil—all that is the movement of becoming: I am this now but I will reach the ultimate destination, the highest principle, whether that principle be called god, Brahman, or any other name. The constant struggle whether to become, or to be, is the same. But when one is trying to become, in various directions, then you are denying being. When you try to be you are becoming also. See this movement of the mind, of thought: I think I am, and being dissatisfied, discontented, I try to fulfil myself in something; I drive towards a particular goal; it may be painful, but the end is thought to be pleasurable. There is this constant struggle to be and to become.

这种检查的要素之一是智慧。智慧是识别、理解、分辨的能力，它也是观察、汇集我们收集的一切并据此行动的能力。那种收集、识别、观察有可能包含成见。当存在成见的时候，智慧就被否定了。如果你追随另一个人，智慧就会被否定。因为对他的追随无论多么高尚，这都会否定你自己的觉察，否定你自己的观察——虽然你只是在追随一个会告诉你要做什么、想什么的人。如果你那样做，那么智慧就不存在，因为其中没有观察，因而也没有智慧。智慧需要怀疑、质疑，而不是被别人的热情和活力打动。智慧需要非个人的观察。它不仅是理解能力，那是理智的、口头上的解释，还包括我们可以尽可能多地收集关于人或事的信息，以及知道那些信息永远不可能是完整的。有智慧的地方就会有疑虑、观察和理性的、非个人化的清晰思考。对整个人类拥有的复杂性，拥有的生理反应、情感反应，智能、爱和痛苦一目了然，是最高的智慧。迄今为止，智慧没能超越冲突。我们将一起看一看，脑子有没有可能使我们免于冲突。从出生到死亡，冲突一直伴随着我们。在所谓的精神上或心理上，我们一直拥有想成为什么、变成什么的内心挣扎，想在世界上获得成功，追求满足——这一切都是“成为”的活动：现在我是这样的，但是我将达成最终的目标、最高的标准，无论那个标准是被称为神、婆罗门还是任何其他名字。无论是成为什么还是变成什么，持续的挣扎都是相同的。但是当你试图在各个方面变成其他样子，你就是在否定此刻的存在。你想要成为什么样子，就会成为什么样子。看看这个内心的、思想的活动：我认为我是什么，感到不满意、不满足，我试图在某件事上满足自己；我朝着特定的目标前进；它或许是痛苦的，但结果被认为是快乐的。这就是存在和成为的持续冲突。

>> We are all trying to become: physically, we want a better house, a better position with more power, higher status. Biologically, if we are not well, we seek to become well. Psychologically, the whole inward process of thought, of consciousness, the whole drive, inwardly, is from the recognition that one is actually nothing, and by becoming, to move away from that. Psychologically, inwardly, there is always the escape from "what is", always the running away from that which I am, from that with which I am dissatisfied to something which will satisfy me. Whether that satisfaction is conceived as deep contentment, happiness, or enlightenment, which is a projection of thought, or as acquiring greater knowledge, it is still the process of becoming—I am, I shall be. That process involves time. The brain is "programmed" to this. Not only in this Western

world but in the East, everyone is trying to become, or to be, or to avoid. Now, is this the cause of conflict, inwardly and outwardly? Inwardly there is this imitation, competition, conformity with the ideal; outwardly there is this competition between so-called individuals of one group against another group, nation against nation. Inwardly and outwardly there is always this drive to become and to be something.

我们都试图成为这样的人：物质上，我们想要更好的房子、更大权力的职位、更高的地位；生理上，如果我们不健康，我们会设法变得健康；心理上，思想意识的全部内在驱力，都出自你认为自己实际上一无是处，并想通过成为更好的自己来逃离现状。在心理上，我们总想逃避“是什么”“我是谁”这样的问题，逃离那些已有的、让我们感到满足的东西，奔向也许会满足我的东西。无论这种满足被认为是深深的满意、幸福或启迪——那是思想的投射，还是能获取更多的知识，都仍然是成为的过程——我是这样的，我将成为那样的。这个过程涉及时间。脑子被设定为这样的。不仅在西方世界，在东方世界也是如此，每个人都在试图成为什么，作为什么，或逃离什么。那么，这是不是内在和外在冲突的原因？内在是模仿、竞争、对理想的遵从；外在是一个所谓的群体和群体之间、国家和国家之间的竞争。内在和外在都存在这种去成为和去做什么的驱动力。

>> We are asking: is this the basic cause of our conflict? Is man doomed—as long as he lives on this marvellous earth—to perpetual conflict? One can rationalize this conflict, say nature is in conflict, the tree struggling to reach the sun is in conflict, and that that is part of our nature, because, through conflict, through competition, we have evolved, we have grown into this marvellous human being that we are—this is not being said sarcastically. Our brain is programmed to conflict. We have a problem which we have never been able to resolve. You may neurotically escape into some fantasy and in that fantasy be totally content, or you may

imagine that you have inwardly achieved something and be totally content with that: an intelligent mind must question all this, it must exercise doubt, scepticism. Why have human beings, for millions of years, from the beginning of man up to the present time, lived in conflict? We have accepted it, we have tolerated it, we have said it is part of our nature to compete, to be aggressive, to imitate, to conform; we have said that it is part of the everlasting pattern of life.

我们要问：这是冲突的基本原因吗？只要人类活在这个不可思议的地球上，就注定要处于无止境的冲突中吗？你可以解释这个冲突，说它是天性，树挣扎着去吸收阳光就是冲突，那是我们本性的一部分，因为通过冲突、通过竞争，我们进化了，我们成长为现在这样非凡的人类——这不是在讽刺。我们的脑子被设定成冲突。我们有一个永远也不曾解决的问题。你可能会神经质地逃入某种幻想并在其中获得完全的满足，或者你可以想象自己取得了一些成就并对它感到完全的满足，但一颗智慧的心必须质疑这一切，它必须保持怀疑。为什么数百万年来，人类从开始存在到现在，就一直活在冲突之中？我们接受它，容忍它，我们说竞争、侵略、模仿、遵从是我们天性的一部分，说它是生命永恒模式的一部分。

>> Why is man, who is so highly sophisticated in one direction, so utterly unintelligent in the other directions? Does conflict end through knowledge—knowledge about oneself, or about the world, knowledge about matter, learning more about society so as to have better organizations and better institutions, acquiring more and more knowledge? Will that solve our human conflict? Or is it that freedom from conflict has nothing whatsoever to do with knowledge?

为什么在某些方面高度发展的人类，在其他方面却如此彻底的愚钝呢？通过获得越来越多的知识——关于你自己、关于世界、关于物质的知识，通过更加了解社会以便拥有更好的组织和机构，冲突会终结吗？那能解决我们人类的冲突吗？还是说，从冲突中解脱和知识没有任何关系？

>> We have a great deal of knowledge about the world, about matter and the universe; we have also a great deal of historical knowledge about ourselves: will that knowledge free the human being from conflict? Or has freedom from conflict nothing to do with analysis, with discovering the various causes and factors of conflict? Will the analytical discovery of the cause, or many causes, free the brain from conflict?—the conflict which we have while we are awake during the daytime and the conflict carried on while we are asleep. We can examine and interpret dreams, we can go into the whole question of why human being dream at all; will that solve conflict? Will the analytical mind analysing very clearly, rationally, sanely into the cause of conflict, end conflict? In analysis the analyser tries to analyse conflict, and in doing so separates himself from conflict—will that solve it? Or is it that freedom has nothing whatsoever to do with any of these processes? If you follow somebody who says "I

will show you the way; I am free from conflict and I will show you the way"—will that help you? This has been the part of the priest, the part of the guru, the part of the so-called enlightened man— "Follow me, I will show you;" or, "I will point out the goal to you." History shows this through millennia upon millennia, and yet man has not been able to solve his deep-rooted conflict.

我们拥有大量关于世界、物质和宇宙的知识，也拥有大量关于我们自身历史的知识，那些知识会让人类从冲突中解脱吗？还是说从冲突中解脱与分析、发现冲突的各种原因和条件没有任何关系？对原因的分析和发现，会让脑子从白天清醒时具有的冲突和在入睡时携带的冲突中解脱出来吗？我们可以检查并解释梦，我们可以探究人类到底为什么做梦这个问题，但那会解决冲突吗？非常清晰、理性、稳妥地分析冲突产生原因的这颗善于分析的心，会终止冲突吗？在分析中，分析者试图分析冲突，在这个过程中他将自己从冲突中分离了——这会解决它吗？

还是说那份自由和这些过程没有任何关系？如果你追随的人说“我将指给你方法，我没有冲突，我会告诉你方法”，那会帮助你吗？他显然担任了牧师的角色、古鲁的角色、所谓觉悟者的角色——“跟我来，我会告诉你”或者“我将为你指出目标”。几千年的历史都在上演这件事，然而人类并没有解决他们根深蒂固的冲突。

>> Let us find out together—not agree, not as an intellectual verbal concept—if there is a perception, an action, that will end conflict, not gradually, but immediately. What are the implications of that? The brain being programmed to conflict is caught in that pattern. We are asking if that pattern can be broken immediately, not gradually. You may think you can break it through drugs, through alcohol, through sex, through different forms of discipline, through handing oneself over to something—man has tried a thousand different ways to escape from this terror of conflict. Now, we are asking: is it possible for a conditioned

brain to break that conditioning immediately? This may be a theoretical, non–actual question. You may say it is impossible, it is just a theory, it is just a wish, a desire, to be free of this conflict. But if you examine the matter rationally, logically, with intelligence, you see that time will not solve this conditioning. The first thing to realize is that there is no psychological tomorrow. If you see actually, not verbally, but deeply in your heart, in your mind, in the very very depths of your being, you will realize that time will not solve this problem. And that means that you have already broken the pattern, you have begun to see cracks in the pattern we have accepted of time as a means of unravelling, breaking up, this programmed brain. Once you see for yourself, clearly, absolutely, irrevocably, that time is not a freeing factor then already you begin to see cracks in the enclosure of the brain. Philosophers and scientists have said: time is a factor of growth, biologically, linguistically, technologically, but they have never enquired into the nature of psychological time. Any enquiry into psychological time implies the whole complex of psychological becoming—I am this, but I will be that; I am unhappy, unfulfilled, desperately lonely but tomorrow will be different. To perceive that time is the factor of conflict then that very perception is action; decision has taken place, the very perception is the action and decision.

让我们一起探寻——不要赞同，不是作为一种智力的语言概念——是否有一种觉察、行动，它将会终结冲突，不是逐渐地而是立即结束。其中的含义是什么？其实就是被训练成充满冲突的脑子陷在了那个模式中。我们要问，那个模式能否立即被打破，而不是逐渐地。你也许认为你可以通过药物、酒精、性、各种形式的训练、把自己交付给什么来打破它——人类已经尝试了上千种不同的方式来逃避这个可怕的冲突了。现在我们要问：对一个被约束的头脑来说，有没有可能立即打破那个约束？这也许是一个理论化的、不实际的问题。你也许会说那是不可能的，从这个冲突中解脱只是一种理论，只是一种愿望、一种欲望。但是如果你有理性地、有逻辑地、带着智慧去检查这个问题，你就会看到，时间不会解决这个约束。首先需要了解的是，不存在心理上的明天。如果你真的去了解，不是口头上，而是在你的内心深处、你的头脑里、你生命的最深处看到这一点，你就会了解，时间不会解决这个问题。你开始看到那个模式的裂隙，即我们接受时间作为拆解、打破这个设定头脑的手段。那意味着你已经打破了那个模式，一旦你清楚地、完全地、不可逆转地看到，时间并不是一个可以解放的因素，那么你就已经看到了头脑围墙之中的裂隙。哲学家和科学家说过，在生物学上、语言学上、技术上，时间都是成长的要素，但他们从来没有探究过心理时间的本质。对心理时间的探究涉及"成为什么"这种心理形成的全部复杂过程——我是这样的，但是我将会是那样的；我不快乐、不满足、极度孤独，但是明天将会不同。觉察到时间是冲突的要素，那份觉察就是行动；决定已经发生，那份觉察正是行动和决定。

>> There are multiple forms of conflict, there are thousands

of opinions so there are thousands of forms of conflict. But we are not talking about the many forms of conflict but about conflict itself. We are not talking about your particular conflict—I don't get on with my wife, or in my business, or this or that—but the conflict of the human brain in its existence. Is there a perception—not born of memory, not born of knowledge—that sees the whole nature and structure of conflict; a perception of that whole? Is there such perception at all—not analytical perception, not intellectual observation of the various types of conflict, not an emotional response to conflict? Is there a perception not of remembrance, which is time, which is thought? Is there a perception which is not of time or thought, which can see the whole nature of conflict, and with that very perception bring about the ending of conflict? Thought is time. Thought is experience, knowledge, put together in the brain as memory. It is the result of time—"I don't know a week ago but I know now." The multiplication of knowledge, the expansion of knowledge, the depth of knowledge, is of time. So thought is time—any psychological movement is time. If I want to go from here to Montreux, if I want to learn a language, if I want to meet somebody at a distant place, time is required. And that same outer process is carried on inwardly—"I am not, I will be". So thought is time. Thought and time are indivisible.

冲突有多种形式。有成千上万种意见，因而有成千上万种冲突。但是我们不是要谈论冲突的多种形式，而是冲突本身。我们不是在谈论你的特定冲突——我不能与妻子和谐共处，或不能应付我的生意，或这或那——而是在谈论人类头脑本身具有的冲突。有没有一种觉察——不是出于记忆，不是出于知识——能够看到冲突的整个本质和结构，一种对冲突整体的觉察？到底有没有这样的觉察，不是对各种类型的冲突的分析式的觉察、理智的观察，也不是对冲突的情感反应？有没有一种觉察，不是来自记忆，而是来自时间和思想？有没有一种觉察，不是来自时间或思想，而是它能看到冲突的全部本质，并凭此带来冲突的终结？思想是时间，思想是经验、知识、头脑中记忆的组合，它是时间的结果——“我一周之前不知道，但是现在我知道”。知识的增加、扩展、深化都是时间的产物，因此思想就是时间——任何心理上的活动都是时间。如果我想从这里去蒙特勒，如果我想学一种语言，如果我想到遥远的地方见人，时间是必需的。同样的外部过程在内心继续——“我不是，我将会是”。因此，思想即是时间，思想和时间是不可分割的。

>> And we are asking the question: is there a perception which is not of time and thought—a perception that is entirely out of the pattern to which the brain has been accustomed? Is there such a thing that perhaps alone is going to solve the problem? We have not solved the problem in a million years of conflict; we are continuing the same pattern. We must find, intelligently, hesitantly, with care, if there is a way, if there is a perception which covers the whole of conflict, a perception which breaks the pattern.

我们的问题是：有没有一种觉察无关时间和思想——一种完全脱离头脑习惯模式的觉察？有没有这样一种或许会独立解决问题的东西？我们在上百万年的冲突中并没有解决问题，我们延续着相同的模式。我们必须睿智地、试探性地、小心翼翼地去发现，是否有一种方法、一种觉察，它能覆盖所有的冲突，并且能够打破模式。

>> The speaker has put this question forward. Now how shall we meet this together? He may be wrong, irrational, but after you have listened to him very carefully, it is your responsibility as well as the speaker's, to see if it is so, if it is possible. Do not say: "Well it is not possible because I have not done it; it is not within my sphere; I have not thought enough about it; or, I do not want to think about it at all because I am satisfied with my conflict and because I am quite certain one day humanity will be free of conflict." That is all just an escape from the problem. So are we together being aware of all the complexities of conflict, not

denying it. It is there, it is there as actually as pain in the body. Are we aware without any choice that it is so and at the same time ask the question as to whether there is a different approach altogether?

讲话者进一步提出了这个问题。那么我们将如何共同面对它？他也许是错误的、荒谬的，但是在你们非常认真地听他说完之后，请看一看是不是这样、是不是可能的。这是你们的责任，也是讲话者的责任。不要说“那是不可能的，因为我没有那样做过；那不在我的兴趣范围内；我没有充分考虑过它；我根本不想考虑它，因为我对我的冲突很满意，而且我相当确定有一天人类会从冲突中解脱出来”，那些都只是对问题的逃避。所以，我们是不是共同意识到了冲突的全部复杂性，而不是在否认它。它就在那儿，和身体的疼痛一样实际。我们是否毫无选择地觉察到它确实是这样的，并同时提出这个问题：是否存在一种完全不同的方式？

>> Now, can we observe—it does not matter what it is—without the naming, without the remembrance? Look at your friend, or your wife, or whomever it is, observe that person without the words "my wife" or "my friend" or "we belong to the same group" without any of that—observe so that you are not observing through remembrance. Have you ever directly tried it? Look at the person without naming, without time and remembrance and also look at yourself—at the image that you have built about yourself, the image that you have built about the other; look as though you were looking for the first time—as you might at a rose for the first time. Learn to look; learn to observe this quality which comes without all the operation of thought. Do not say it is not possible. If you go to a professor, not knowing his subject but wanting to learn from him (I am not your professor), you go to listen. You do not say: "I know something about it," or "You are wrong," or "You are right," or "I don't like your attitude." You listen, you find out. As you begin to listen sensitively, with awareness,

you begin to discover whether it is a phoney professor using a lot of words, or a professor who has really gone into the depths of his subject. Now, can we together so listen and observe, without the word, without remembrance, without all the movement of thought? Which means, complete attention; attention, not from a centre but attention which has no centre. If you have a centre from which you are attending, that is merely a form of concentration. But if you are attending and there is no centre, it means that you are giving complete attention; in that attention there is no time.

那么，我们能否不去定义、不带记忆地观察，无论你观察的是什么？看看你的朋友、你的妻子或无论什么人，观察那个人而不带着“我的妻子”“我的朋友”“我们属于同一个团体”这样的想法——什么都不带地去观察，但不是通过记忆观察。你曾经直接地尝试过吗？抛掉名称、时间和记忆地去看别人，也看你自己——看你建立的关于自己的形象、关于别人的形象，如同你是初次看一样——就像你第一次看一株玫瑰一样。请学习看，学习观察这种品质，它来自彻底没有思想的运作，别说那是不可能的。如果你去找一个教授，你不知道他研究的主题，但是想向他学习（我不是你的教授），你就会去听。你不会说“我知道它”“你是错的”“你是对的”，或“我不喜欢你的态度”。你会去听，去了解。当你开始带着觉察和敏感去倾听，你会开始发现他到底是一个喋喋不休的、假冒的教授，还是一个真正对他的主题有深入研究的教授。现在，我们能不能这样共同来倾听和观察，不带名称、不带记忆、不带一切思想活动？那意味着完全的注意。这种注意不是来自一个中心，而是没有中心。如果你有一个中心，你从那个中心出发去注意，那就只是一种专注的形式。但是如果你的注意没有中心，那意味着你是完全的注意，在那注意中，不存在时间的概念。

>> Many of you, fortunately or unfortunately, have heard the speaker for many years and one sees that this breaking of the "programme" of the brain has not come about. You

repeatedly listen to that statement year after year and it has not come about. Is it because you want to attain, to become, to have that state in which the pattern of the brain has been broken? You have listened, and it has not come about, and you are hoping that it will come about—which is another form of striving to become. So you are still in conflict. So you brush it all aside and say you will not come here any more because you have not got what you want—"I want that but have not got it." That wanting is the desire to be something and is a cause of conflict. That desire comes from the "programmed" brain. We are saying: to break that programme, that pattern, observe without the movement of thought. It sounds very simple, but see the logic of it, the reason, the sanity of it, not because the speaker says so, but because it is sane. Obviously one must exercise the capacity to be logical, rational and yet know its limitation; because rational, logical thinking is still part of thought. Know that thought is limited, be aware of that limitation and do not push it further because it will still be limited however far you go, whereas if you observe a rose, a flower, without the word, without naming the colour, but just look at it, then that look brings about great sensitivity, breaks down this sense of heaviness of the brain, and gives extraordinary vitality. There is a totally different kind of energy when there is pure perception, which is not related to thought and time.

16th July, 1981

你们很多人有幸或不幸地听讲话者谈了很多年，而打破头脑“程序”的事情并没有发生。你们年复一年地反复听那个陈述，但它并未发生。是你想达到、想成为、想拥有的头脑模式被打破的那种状态导致的吗？你听了，它没发生，你希望它发生，这是另一种努力，因此仍然在冲突中。于是你把这一切丢开，说你不会再来这里了，因为你并未得到你想要的东西——“我想要那个但我并没有得到”。那个想要成为什么的欲望，就是冲突的原因。那个欲望就来自“程序化”的头脑。我们要说的就是，打破那个程序、那个模式，没有思想活动地去观察。这听上去很简单，但是要看到它的逻辑、它的原因、它的合理性，不是因为讲话者这么说，而是因为它合乎情理。显然一个人必须训练自己的逻辑能力和理性思维而又知道它的局限，因为理性、逻辑地思考仍然是思想的一部分。知道思想是局限的，觉察那个局限，不要让它走得太远。因为无论你走多远，它仍将是局限的。然而如果你观察一朵花，如一株玫瑰，但没有评价，不去命名它的颜色，而只是看着它，那么那个看就会带来巨大的敏感，就会弱化头脑的沉重感，并且带来非凡的活力。当纯粹的觉察发生的时候，就会产生一种全然不同的能量，它跟思想和时间无关。

一九八一年七月十六日

>> Order is necessary in our everyday activity; order in our action and order in our relationship with each other. One has to understand that the very quality of order is totally different from that of discipline. Order comes through

directly learning about ourselves—not according to some philosophers or some psychologists. We discover order for ourselves when we are free from all sense of compulsion, from all sense of determined effort to obtain order along a particular path. That order comes very naturally. In that order there is righteousness. It is order, not according to some pattern, and not only in the outward world, which has become so utterly chaotic, but inwardly within ourselves where we are not clear, where we are confused and uncertain. Learning about ourselves is part of order. If you follow another, however erudite, you will not be able to understand yourself.

在我们的日常活动当中，秩序是必要的，包括我们行动中的秩序，以及我们彼此关系中的秩序。你必须了解，秩序的性质完全不同于纪律。秩序来自对我们自己的直接了解，而不是依照某些哲学家或心理学家的说法。当我们从所有的强迫感、所有沿着特定道路去达成秩序的坚定努力中解脱出来的时候，我们就会亲自发现秩序是什么。那秩序会非常自然地到来。在那样的秩序中就有正直。这种秩序不是依照某种模式，它已经彻底变得混乱，也不是只存在于外部世界，而是在我们的内在，在那里我们不清晰、困惑、不确定。了解我们自己，是秩序的一部分。而如果你追随另一个人，无论他多么渊博，你都不会了解自己。

>> To find out what order is we must begin to understand the nature of our relationships. Our life is a movement in relationship; however much one may think one lives alone, one is always related to something or other, either to the past or to some projected image in the future. So, life is a movement in relationship and in that relationship there is disorder. We must examine closely why we live in such disorder in our relationships with each other—however intimate or superficial.

要弄清秩序是什么，我们必须开始了解我们之间关系的本质。我们的生活是一种关系的运动，无论你在何种程度上认为自己是独立的，你都和其他事物或人有关，或者是和过去有关，或者是与将来的某个投影有关。因此，生活是一种关系中的运动，在那个关系中存在着无序。我们必须在我们和他人的关系中——无论是亲密的还是肤浅的——严密检视，我们为什么活在这样的无序当中。

>> The speaker is not trying to persuade you to think in a particular direction, or put any kind of persuasive, subtle pressure on you. On the contrary, we are together thinking over our human problems and discovering what our relationship with each other is and whether in that relationship we can bring about order. To understand the full meaning of relationship with each other, however close, however distant, we must begin to understand why the brain creates images. We have images about ourselves and images about others. Why is it that each one has a peculiar image and identifies himself with that image? Is the image necessary, does it give one a sense of security? Does not the

image bring about the separation of human beings?

讲话者不是试图劝说你在一个特定的方向上思考，或者向你施加任何说服力和微妙的压力。相反，我们是在一同考虑我们人类的问题，发现我们相互之间的关系是什么，以及我们能否在那份关系当中带来秩序。要理解彼此间关系的全部意义，无论这种关系多么亲密或疏远，我们都必须开始去了解脑子为什么会制造形象。我们有关于我们自己的形象，也有关于其他人的形象。为什么每个人都有一个特殊的形象并习惯将自己和它联系起来？形象是必要的吗？它会为一个人带来安全感吗？难道形象不会带来人类的分裂吗？

>> We have to look closely at our relationship with wife, husband or friend; look very closely, not trying to avoid it, not trying to brush it aside. We must together examine and find out why human beings throughout the world have this extraordinary machinery that creates images, symbols,

patterns. Is it because in those patterns, symbols and images, great security is found?

我们必须严密审视我们同妻子或丈夫或朋友之间的关系，非常密切地看，不要试图逃避它、对它视而不见。我们必须一同检查并发现，为什么全世界的人都有这种奇特的制造形象、符号、模式的机制。是因为在那些模式、符号和形象中能找到巨大的安全感吗？

>> If you observe, you will see that you have an image about yourself, either an image of conceit which is arrogant, or the contrary to that. Or you have accumulated a great deal of experience, acquired a great deal of knowledge, which in itself creates the image, the image of the expert. Why do we have images about ourselves? Those images separate people. If you have an image of yourself as Swiss or British or French and so on, that image not only distorts your observation of humanity, but also separates you from others. And wherever there is separation, division, there must be

conflict—as there is conflict going on all over the world, Why? Is it because of our education, because of our culture in which the individual is the most important and where the collective society is something totally different from the individual? That is part of our culture, part of our daily education. When one has an image about oneself as being British or American, that image gives one a certain security. That is fairly obvious. Having created the image about oneself that image becomes semi-permanent; behind that image, or in that image, one tries to find security, safety, a form of resistance. When one is related to another, however delicately, however subtly, psychically or physically, there is a response based on an image. If one is married or related intimately with somebody, an image is formed in one's daily life; whether one is acquainted for a week or ten years, the image is slowly formed about the other person step by step; every reaction is remembered, adding to the image and stored up in the brain so that the relationship—it may be physical, sexual, or psychical—is actually between two images, one's own and the other's.

如果去观察，你就会看到，你有一个关于自己的形象，要么傲慢自负，要么和它相反。或者你积累了大量的经验，获得了大量的知识，因而就制造了专家的形象。我们为什么会有关于自身的形象？那些形象使人们有了区别。如果你把自己想象成瑞士人、英国人或法国人等等，这个形象不但扭曲了你对人性的观察，而且还把你和其他人区分开来。有分裂的地方必定会有冲突——整个世界都在发生冲突。为什么？是因为我们的教育吗？是因为我们个人至上的文化吗？在那里集体和个体是完全不同的东西。那是我们文化的一部分，我们日常教育的一部分。当你把自己想象成是英国人或美国人时，这个形象会带给你某种安全感，这相当明显。制造了关于自己的形象，那个形象就会变成半永久的。你试图在那个形象背后寻找保证、安全，因为它是一种抗拒的形式。当你和另一个人有关的时候，无论多么精致、微妙，精神上或身体上都有一种基于形象的反应。如果你与某人结婚，或处于亲密关系中，一个形象就在你的日常生活中建立了。无论你们认识一个星期还是十年，你对另一个人的形象都会逐渐地建立。每一个反应都会被记住，添加到图像中，并储存在脑子里，它可能是身体的、性的或心灵的关系，实际上都是两个形象之间的关系，是你自己的形象和另一个人的形象之间的关系。

>> The speaker is not saying something extravagant, or exotic, or fantastic, he is merely pointing out that these images exist. These images exist and one can never know another completely. If one is married or one has a girlfriend, one can never know her completely; one thinks one knows her because having lived with that person one has accumulated memories of various incidents, various irritations and all the occurrences which happen in daily life; and she also has experienced her reactions and their images are established in her brain. Those images play an extraordinarily important part in one's life. Apparently very few of us are free from

any form of image. The freedom from images is real freedom. In that freedom there is no division brought about by images. One lives with that complex of images, which is one's conditioning. And however much one may talk about brotherhood, unity, wholeness, it is merely empty words having no actual daily meaning. But if one frees oneself from all that imposition, all the conditioning of all that superstitious nonsense, then one is breaking down the image. And also in one's relationship, if one is married or lives with somebody, is it possible not to create an image at all—not to record an incident which may be pleasurable or painful, in that particular relationship, not to record either the insult or the flattery, the encouragement or discouragement?

讲话者并不是在谈论大而无当、稀奇古怪或异想天开的东西，他只是指出这些形象的存在。这些形象是存在的，一个人永远也无法完全了解另一个人。如果你结婚了或者你有一个女朋友，你永远也不能完全了解她；你认为你了解她，因为你和她一同生活过，积累了各种事件、各种烦恼，以及拥有日常生活中所有发生过的事情的记忆；她也发现了她的反应，并在她的头脑中建立起它们的形象。这些形象在一个人的生活中扮演了异常重要的角色。显然，我们很少有人能从形象中解脱。从形象中解脱的自由是真正的自由，在那份自由中，没有形象带来的分裂。你和那些复杂的形象生活在一起，它就是你的制约。无论你怎么谈论兄弟情谊、团结和整体，那只是没有任何实际意义的空话。但是如果你将自己从所有那些强加的东西、所有那些迷信的胡说八道带来的制约中解脱出来，那么你就打破了形象。如果你结婚了或者和某人一起生活，有没有可能在你们的关系中完全不制造形象——不记录一件令人快乐或痛苦的事情，在那种特定的关系中，不记录侮辱，也不记录奉承、鼓励或沮丧。

>> Is it possible not to record at all? Because if the brain

is constantly recording everything that is happening, psychologically, then it is never free to be quiet, it can never be tranquil, peaceful. If the machinery of the brain is operating all the time, it wears itself out. This is obvious. It is what happens in our relationships with each other—whatever the relationship is—and if there is constant recording of everything then the brain slowly begins to wither away and that is essentially old age.

有没有可能完全不记录？因为如果脑子不停地记录心理上发生的每件事情，它就永远也不会有安静的自由，它永远也不可能平静、平和。如果头脑机制一直在运作，那么它就会让自己疲惫不堪，这是很明显的。这是在我们彼此的关系中发生的事情——无论这个关系是怎样的——如果存在对每件事情的持续的记录，那么头脑就会慢慢地开始干枯，那是本质上的衰老。

>> So in investigating we come upon this question: is it possible

in our relationships with all their reactions and subtleties, with all their essential responses, is there a possibility of not remembering? This remembering and recording is going on all the time. We are asking whether it is possible not to record psychologically, but only to record that which is absolutely necessary? In certain directions it is necessary to record. For example, one must record all that which is necessary to learn mathematics. If I am to be an engineer, I must record all the mathematics related to structures and so on. If I am to be a physicist, I must record that which has already been established in that subject. To learn to drive a car I must record. But is it necessary in our relationships to record, psychologically, inwardly, at all? The remembrance of incidents past, is that love? When I say to my wife, "I love you," is that from a remembrance of all the things we have been through together—the incidents, the travail, the struggles, which are recorded, stored in the brain—is that remembrance actual love?

于是在调查中我们遇到了这个问题：有没有可能不去记住在我们的关系中，伴随产生的所有的反应和微妙之处？尽管这种记住和记录一直都在继续。我们要问，有没有可能在心理上不是什么都去记录，而只是记录那些绝对必要的东西？在某些方面记录是必要的。例如，你必须记录所有那些对学习数学来说是必要的东西。如果我打算成为一名工程师，我必须记录所有和结构等相关的数学。如果我打算成为一个物理学家，我也必须记录那些在这个学科中已经确立的东西。要学会驾驶汽车，我也必须记录。但是在我们的关系当中，在心理上，到底有没有记录的必要？对过去的事情的记忆，那是爱吗？当我对我的妻子说“我爱你”，那是来自对我们共同经历过的所有事情的记忆——事件、艰苦、奋斗，那些被记录、储存在头脑里的东西——那些记忆是真正的爱吗？

>> So is it possible to be free and not to record psychologically at all? It is only possible when there is complete attention. When there is complete attention there is no recording.

因此，我们到底有没有可能成为自由的人，不在心中进行任何记录？其实只有完整的注意存在时，这种自由才是可能的。因为当存在完整的注意时，就不存在记录。

>> I do not know why we want explanations, or why it is that our brains are not swift enough to capture, to have an insight into, the whole thing immediately. Why is it that we cannot see this thing, the truth of all this, and let that truth operate and therefore cleanse the slate and have a brain that is not recording at all psychologically? But most human beings are rather sluggish, they rather like to live in their old patterns, in their particular habits of thought; they reject anything new because they think it is much better to live with the known rather than with the unknown. In the known there is safety—at least they think there is safety, security—so they keep on repeating, working and struggling within that field of the known. Can we observe without the whole process and machinery of memory operating?

我不知道为什么我们总想解释，又或者为什么我们的脑子不够敏捷，无法迅速地捕捉到对整个事情的洞察。为什么我们不能看清这件事情，看到这一切的真相，并让那个真相去运作，从而清理那个记录，并拥有一个完全不去记录的脑子？但是大多数人宁愿没精打采，他们宁愿活在自己陈旧的生活模式里，活在他们特定的思想习惯里，任何新东西他们都拒绝，因为他们认为活在已知里比活在未知里更好。已知中有安全感——至少他们认为有安全、保障——因此，他们继续在已知的范围内重复着、工作着、挣扎着。我们能在没有记忆运作的程序和机制的状态下观察吗？

>> What is love? This is a very complex question; all of us feel we love something or other, abstract love, love of a nation, love of a person, love of gardening, love of overeating. We have abused the word love so greatly that we have to find out basically what love is. Love is not an idea. Love of god is an idea, love of a symbol is still an idea. When you

go to the church and kneel down and pray, you are really worshipping, or praying to, something which thought has created. So, see what is happening, thought has created it—actually this is a fact—and you worship that which thought has created; which means you are worshipping, in a very subtle way, yourself. This may seem a sacrilegious statement, but it is a fact. That is what is happening throughout the world. Thought creates the symbol with all the attributes of that symbol, romantic or logical and sane; having created it you love it, you become totally intolerant of any other thing. All the gurus, all the priests, all the religious structures, are based on that. See the tragedy of it. Thought creates the flag, the symbol of a particular country, then you fight for it, you kill each other for it, and so the flag becomes a symbol of your love. We have lived for millions of years that way.

什么是爱？这是一个非常复杂的问题。我们所有人都觉得我们爱着什么，抽象的爱，对国家的爱，对一个人的爱，对园艺的爱，对暴饮暴食的爱。我们对“爱”这个字的滥用是如此严重，因而我们必须从根本上搞清楚什么是爱。爱不是观念。对神明的爱是一个观念，对符号的爱仍然是一个观念。当你在教堂跪下来祷告，你实际上是在崇拜或祈求某个思想制造的东西。因此，看看正在发生的是什么，思想制造了它——这是一个事实——而你崇拜那个思想制造的东西，那意味着你正在以一种非常微妙的方式崇拜你自己。这看上去像是一种亵渎的陈述，但这是事实。这是全世界都在发生的事情。思想制造了符号以及有关那个符号的所有浪漫的、合理的、明智的特质。通过制造它，你爱上它，并变得完全不能容忍其他东西。所有的古鲁、牧师和宗教机构都建立在那上面，请看到它的悲惨。思想制造了一个国家的特殊旗帜、符号，然后你们为它而战，为它相互杀戮，因此旗帜变成了你爱的符号。我们已经这样生活了数百万年。

>> When we say we love another, in that love there is desire, the pleasurable projections of the various activities of thought. One has to find out whether love is desire, whether love is pleasure, whether in love there is fear; for where there is fear there must be hatred, jealousy, anxiety, possessiveness, domination. There is beauty in relationship and the whole cosmos is a movement in relationship. Cosmos is an order and when one has the order in oneself, one has the order in one's relationships and therefore the possibility of the order in our society. If one enquires into the nature of relationship, one finds it is absolutely necessary to have order, and out of that order comes love. What is beauty? You see the fresh snow on the mountains this morning, clean, a lovely sight. You see those solitary trees standing black against that white. Looking at the world about us you see the marvellous machinery, the extraordinary computer with its special beauty; you see the beauty of a face, the beauty of a painting, the beauty of a poem—you seem to recognize beauty out there. In the museums or when

you go to a concert and listen to Beethoven, or Mozart, there is great beauty—but always out there. In the hills, in the valleys with their running waters, and the flight of birds and the singing of a blackbird in the early morning, there is beauty. But is beauty only out there? Or is beauty something that only exists when the "me" is not? When you look at those mountains on a sunny morning, sparkling clear against the blue sky, their very majesty drives away all the accumulated memories of yourself—for a moment. There the outward beauty, the outward magnificence, the majesty and the strength of the mountains, wipes away all your problems—if only for a second. You have forgotten yourself. When there is total absence of yourself beauty is. But we are not free of ourselves; we are selfish people, concerned with ourselves, with our importance or with our problems, with our agonies, sorrows and loneliness. Out of desperate loneliness we want identification with something or other and we cling to an idea, to a belief, to a person, especially to a person. In dependency all our problems arise. Where there is psychological dependency, fear begins. When you are tied to something, corruption begins.

当我们说我们爱另一个人，那爱中有着欲望，有着各种思想活动对快乐的投射。你必须弄清楚爱是不是欲望、是不是快乐，其中有没有恐惧，因为有恐惧的地方必定会有憎恨、嫉妒、焦虑、占有和控制。关系中存在着美，整个宇宙就是一种关系中的运动。宇宙是有秩序的，当你心里有了秩序，你的关系中就有了秩序，因而秩序就会在我们的社会中成为可能。如果探究关系的本质，你就会发现，拥有秩序是绝对必要的，从那份秩序中就会产生爱。美是什么？今天早上你看到山上的新雪洁白迷人，你看到那些独自矗立的黑色树木和白雪相互映衬。看看我们周围的世界，你会看到不可思议的机器、非凡的计算机以及它特殊的美，你看到一张脸孔的美、一幅画的美、一首诗的美——你似乎认出了外在的美。在博物馆里，或当你去剧院听贝多芬或莫扎特音乐的时候，就能感受到巨大的美——但这是外在的。山间、峡谷中奔腾的流水中，鸟儿的飞翔中，清晨黑鸟的歌唱中，都存在着美。但是美只是外在的吗？还是说只有在“我”不存在的时候美才会出现？一个阳光明媚的早晨，你看着群山和晶莹剔透的蓝天，正是它们的庄严驱散了你自己积累的记忆——在一瞬间。在那里，外在的美，外在的壮丽，群山的庄严和力量将你的一切问题一扫而光——只要一瞬间。你忘了你自己。当你彻底不在的时候，美就在了。但是我们不能摆脱自身，我们是自私的人，只关心自己，关心我们的重要性或我们的问题、痛苦、伤悲和孤独。因为令人绝望的孤独，我们想要认同某件事，我们紧紧抓住一个想法、一个信念、一个人，尤其是一个人。在依赖当中，我们所有的问题都出现了。当有心理上的依赖时，恐惧就开始了。当你被某种东西束缚的时候，腐败就开始了。

>> Desire is the most urgent and vital drive in our life. We are talking about desire itself, not desire for a particular thing. Substitute an image for the actual. The actual is desire—the burning of it and they think that one can overcome that

desire by substituting something else for it. Or, surrender yourself to that which you think is the master, the saviour, the guru—which again is the activity of thought. One has to understand the whole movement of desire; for obviously it is not love, nor yet compassion. Without love and compassion, meditation is utterly meaningless. Love and compassion have their own intelligence which is not the intelligence of cunning thought.

欲望是我们生命中最为急迫和活跃的驱动力。我们谈论的是欲望本身，不是对特定事物的欲望。不要用一个形象代替真实的东西。实际存在的是欲望，它在燃烧，人们却认为一个人可以通过用别的东西替代它来克服那个欲望，或者把自己交给那个他认为是大师、救世主、古鲁的人，但那还是思想的活动。你必须了解欲望的整个运动，因为显然它不是爱，也不是激情。没有爱和激情，冥想就完全没有意义。爱和激情有它们自身的智慧，而智慧不属于狡猾的思想。

>> So it is important to understand the nature of desire, why it has played such an extraordinarily important part in our life; how it distorts clarity, how it prevents the extraordinary quality of love. It is important that we understand and do not suppress, do not try to control it or direct it in a particular direction, which you think may give you peace.

因此，重要的是了解欲望的本质，它为什么在我们的生命中扮演了如此重要的角色，它如何扭曲了事实，如何阻碍了爱的非凡品质。重要的是我们去了解它，不是压抑，不是试图控制它，或在一个你认为或许会带给你和平的特定方向上引领它。

>> Please bear in mind that the speaker is not trying to impress you or guide and help you. But together we are walking a very subtle, complex path. We have to listen to each other to find out the truth about desire. When one understands the significance, the meaning, the fullness, the truth of desire, then desire has quite a different value or drive in one's life.

请记住，讲话者不是要打动、指引或帮助你，而是我们一同走在一条非常微妙、复杂的路上。我们必须彼此倾听，来发现有关欲望的真相。当你了解欲望的意义、内涵和真相的时候，它在我们的生活中就有了完全不同的价值或动力。

>> When one observes desire, is one observing it as an outsider looking at desire? Or is one observing desire as it arises? Not desire as something separate from oneself, one is desire. You see the difference? Either one observes desire, which one has when one sees something in the shop window which pleases one, and one has the desire to buy it so that the object is different from "me", or else the desire is "me", so there is a perception of desire without the observer watching desire.

当你观察欲望的时候，你是作为一个局外人在看那个欲望，还是说在欲望升起的时候你才开始注意它？欲望不是和自己分离的东西，你就是欲望本身。你看到其中的不同了吗？当你去观察欲望，当你在商店橱窗前看到让你产生喜悦之情的东西时，你产生了购买它的欲望，这和你本身就有购买它的欲望是不同的。商品给你带来的欲望和你本身的欲望是不同的。当你不再从观察者的高度注视欲望，你才能觉察欲望；要么，欲望就是“我”。因此，只有对欲望的觉察，而没有注视着欲望的观察者。

>> One can look at a tree. “Tree” is the word by which one recognizes that which is standing in the field. But one knows that the word “tree” is not the tree. Similarly one’s wife is not the word. But one has made the word one’s wife. I do not know if you see all the subtleties of this. One must very clearly understand, from the beginning, that the word is not the thing. The word “desire” is not the feeling of it—the extraordinary feeling there is behind that reaction. So one must be very watchful that one is not caught in the word. Also the brain must be active enough to see that the object may create desire—desire which is separate from the

object. Is one aware that the word is not the thing and that desire is not separate from the observer who is watching desire? Is one aware that the object may create desire but the desire is independent of the object?

你能够看一棵树。“树”是一个字，通过它你去识别那个矗立在地上的东西。但是你知道“树”这个字并不是树。类似地，你的妻子也不是那个词，但是你制造了“妻子”这个词。我不知道你是否看到了其中的微妙。从一开始你就必须非常清楚，词语不是事物本身。“欲望”这个词不是那个感受——那个反应背后有着不同寻常的感受。因此，你必须非常警惕自己有没有陷入词句当中。头脑也必须足够敏锐，能够看到被注视的对象会制造欲望——和对象分离的欲望。你意识到词语不是事物本身了吗？你意识到欲望和那个注视欲望的观察者并不是分离的吗？以及是否意识到对象可以制造欲望，但是欲望并不依赖于那个对象？

>> How does desire flower? Why is there such extraordinary energy behind it? If we do not understand deeply the nature of desire, we will always be in conflict with each other. One may desire one thing and one's wife may desire another and the children may desire something different. So we are always at loggerheads with each other. And this battle, this struggle, is called love, relationship.

欲望是如何成熟的？为什么它背后会有如此非同寻常的活力？如果没有深刻理解欲望的本质，我们就会永远处于相互的冲突之中。你可能对一个东西有欲望，而你的妻子可能想要另一个东西，孩子们想要的或许也不同，因此我们总是处于彼此的争执中。而这种争斗、挣扎被称为爱、关系。

>> We are asking: what is the source of desire? We must be very truthful in this, very honest, for desire is very very deceptive, very subtle, unless we understand the root of it. For all of us sensory responses are important—sight, touch,

taste, smell, hearing. And a particular sensory response may for some of us be more important than the other responses. If we are artistic, we see things in a special way. If we are trained as an engineer, then the sensory responses are different, so we never observe totally, with all the sensory responses. We each respond somewhat specially, divided. Is it possible to respond totally with all one's senses? See the importance of that. If one responds totally with all one's senses, there is the elimination of the centralized observer. But when one responds to a particular thing in a special way then the division begins. Find out when you leave this tent, when you look at the flowing waters of the river, the light sparkling on the swiftness of the waters, find out if you can look at it with all your senses. Do not ask me how, for that becomes mechanical. But educate yourself in the understanding of total sensory response.

我们要问：欲望的根源是什么？对这个问题我们必须非常老实、非常诚恳，因为欲望是非常有迷惑性的、非常微妙的，除非我们了解它的根源。对我们所有人来说，感官反应都是重要的——视觉、触觉、味觉、嗅觉、听觉。对我们中的一些人来说，一种特定的感官反应可能比其他的反应更为重要。如果我们具有艺术气质，我们就会用一种特殊的方式看待事物。如果我们被训练成为一名工程师，那么感官反应也会不同，因此我们从来没有用我们所有的感官去完全地观察。我们每个人的反应都存在某种程度的独特和分裂。可不可能用你所有的感官完全地去反应？要看到这件事的重要性。如果你用所有感官完全地去反应，就会排除那个作为中心的观察者。但是当你以一种特定的方式对一件特别的事情做出反应的时候，分裂就开始了。当你离开这个帐篷，当你看到河中的流水、看到急流中闪耀的光芒的时候，去发现你是否能够用全部的感官去看它。不要问我怎么看，因为那会变得机械；而是要在对完全的感官反应的了解中教育你自己。

>> When you see something, the seeing brings about a response. You see a green shirt, or a green dress, the seeing

awakens the response. Then contact takes place. Then from contact thought creates the image of you in that shirt or dress, then the desire arises. Or you see a car in the road, it has nice lines, it is highly polished and there is plenty of power behind it. Then you go around it, examine the engine. Then thought creates the image of you getting into the car and starting the engine, putting your foot down and driving it. So does desire begin and the source of desire is thought creating the image, up to that point there is no desire. There are the sensory responses, which are normal, but then thought creates the image and from that moment desire begins. Now, is it possible for thought not to arise and create the image? This is learning about desire, which in itself is discipline. Learning about desire is discipline, not the controlling of it. If you really learn about something, it is finished. But if you say you must control desire, then you are in a totally different field altogether. When you see the whole of this movement, you will find that thought with its image will not interfere; you will only see, have the sensation and what is wrong with that?

当你看到什么的时候，就会唤起一个反应。你看到一件绿衬衫或者绿裙子，就唤醒了反应，然后接触就发生了。接着思想从接触中制造了你对那件衬衫或裙子产生反应的形象，继而欲望就升起了。或者你在路上看到一辆汽车，它有着漂亮的流线，车身锃亮，马力强劲。然后你围着它转，检查引擎。接着思想制造了你上车、启动、踏上你的脚驾驶它的形象。欲望就是这样开始的，欲望的根源就是制造形象的思想，一直到那个没有欲望的点。感官反应是正常的，但是之后思想制造了形象，从那一刻欲望就开始了。那么，思想有没有可能不出现并制造形象？这是在了解欲望，它本质上就是纪律。了解欲望是纪律，对它的控制则不是。如果你真正去了解欲望，它就结束了。但是如果你说你必须控制欲望，那么你就处在一个完全不同的领域中了。当你看到这个活动的全部，你就会发现思想及其形象不再介入了，你只是看，仍然拥有你的感知，这有什么问题呢？

>> We are all so crazy about desire, we want to fulfil ourselves through desire. But we do not see what havoc it creates in the world—the desire for individual security, for individual attainment, success, power, prestige. We do not feel that we are totally responsible for everything we do. If one understands desire, the nature of it, then where can we find it? Has it any place where there is love? Is love then something so extraordinarily outside of human existence that it has actually no value at all? Or, is it that we are not seeing the beauty and the depth, the greatness and sacredness of the actuality of it; is it that we have not the energy, the time to study, to educate ourselves, to understand what it is? Without love and compassion with its intelligence, meditation has very little meaning. Without that perfume that which is eternal can never be found. And that is why it is important to put the "house" of our life, of our being, of our struggles, into complete order.

19th July, 1981

我们对欲望都是如此疯狂，我们想要通过欲望填满自己。但是我们没有看到它——个人对安全的欲望，对成就、成功、权力、名望的欲望——给世界带来了什么破坏。我们不觉得我们对自己做的每件事情负有完全的责任。如果你了解欲望，了解它的本质，那么它还有什么位置？它在爱中有位置吗？爱是某种如此离奇的超乎人类存在的东西，以至于它实际上完全没有价值吗？还是说我们没有看到它事实上的美、深度、崇高和神圣？我们没有精力、时间去学习、教育自己，去了解它是什么吗？没有爱、慈悲及其智慧，冥想就没有什么意义。没有那种芬芳，你就永远不可能发现那永恒的东西。这就是让承载我们生命和挣扎的“房子”处于完全的秩序中是那么重要的原因。

一九八一年七月十九日

第三章 痛苦和悲伤的终结

The End of Pain and Sorrow

>> We have to consider together whether the brain, which is now only operating partially, has the capacity to function wholly, completely. Now we are only using a part of it, which one can observe for oneself. One can see that specialization, which may be necessary, brings about the functioning of only a part of the brain. If one is a scientist, specializing in that subject, naturally only one part of the brain is functioning; if one is a mathematician, it is the same. In the modern world one has to specialize, and we are asking whether, even so, it is possible to allow the brain to operate wholly, completely.

我们必须一同思考，这个现在只能部分运转的脑子，有没有能力完全地、充分地运作。现在我们只是在使用它的一部分，这点你可以自己观察。你能看到专业化分工——它或许是必要的——导致了脑子只有一部分在运转。如果你是一个科学家，专攻某个学科，自然只有一部分脑子在运作，数学家也一样。在现代社会，一个人必须有所专攻，我们要问的是，尽管如此，有没有可能让脑子完全地、充分地运作。

>> And another question we are asking is: what is going to happen to humanity, to all of us, when the computer outthinks man in accuracy and rapidity—as the computer experts are saying it will? With the development of the robot, man will only have, perhaps, two hours of work a day. This may be going to happen within the foreseeable future. Then what will man do? Is he going to be absorbed in the field of entertainment? That is already taking place: sports are becoming more important; there is the watching of television. Or is he going to turn inwardly, which is

not an entertainment but something which demands great capacity of observation, examination and non-personal perception? These are two possibilities. The basic content of our human consciousness is the pursuit of pleasure and the avoidance of fear. Is humanity increasingly going to follow entertainment? One hopes these gatherings are not a form of entertainment.

我们要问的另一个问题是：当计算机就像专家预言的那样，在准确和速度方面超过人类的时候，对于我们所有人，将会发生什么？随着机器人的发展，人类或许每天只要工作两小时。这可能会在可预见的将来发生。那时人将做什么？他会专注在娱乐中吗？这已经在发生了，运动正在变得越来越重要，还有看电视。还是说他会转向内在？那不是一种娱乐，那需要巨大的观察、检视和非个人化的觉察能力。这是两种可能性。我们人类的基本意识就是追逐快乐和逃避恐惧。人类会越来越多地投入娱乐中吗？希望这些聚会不是一种娱乐。

>> Now, can the brain be totally free so as to function wholly?—Because any specialization, any following of a certain path, a certain groove or pattern, inevitably implies that the brain is functioning partially and therefore with limited energy. We live in a society of specialization—engineers, physicists, surgeons, carpenters and the specializations of particular beliefs, dogmas and rituals. Certain specializations are necessary, but in spite of that can the brain function completely, wholly, and therefore have tremendous energy? This is, I think, a very serious question into which we have to enquire together.

那么，脑子能不能彻底自由，从而完全地运作？因为任何专业化，追随任何特定的道路、规范或模式都必然意味着脑子的局部运作，因而活力是受限的。我们生活在一个专业化的社会里，有着工程师、物理学家、外科医生、木匠，以及专门的特定信仰、教条和仪式。尽管专家是不可缺少的，脑子能不能彻底、完全地运作，因而拥有巨大的活力呢？我认为这是一个非常严肃的问题，我们必须一同来探究。

>> If one observes one's own activity, one finds that the brain functions very partially, fragmentarily, with the result that one's energy becomes less and less as one grows older. Biologically, physically, when one is young one is full of vitality; but as one is educated, and then follows a livelihood that needs specialization, the activity of the brain becomes narrowed down, limited and its energy becomes less and less.

如果观察自己的活动，你会发现脑子的运作是局部的且是支离破碎的，伴随的结果是，随着年龄的增长，你的精力变得越来越少。在生物意义上，在生理上，你在年轻的时候充满了活力，接受过教育，然后从事一种需要专业化的营生，脑子的活动开始窄化、受限，它的活力变得越来越低。

>> Though the brain may have to have a certain form of specialization—as a surgeon for example, can it also operate wholly? It can only operate wholly, with all the tremendous vitality of a million years behind it, when it is completely free. Specialization, which is now necessary for a livelihood, may not be necessary if the computer takes over. It will not take over surgery, obviously. It will not take over the feeling of beauty, as when looking at the evening stars, but it may take over other functions altogether.

尽管脑子或许需要某种形式的专业化，例如作为外科医生，但它是否还能够完全地运作？只有在它完全自由的时候，它落后了一百万年的所有惊人的活力才能完全地运作。专业化对于谋生来说目前是必要的，如果被计算机接管或许就不必要了。显然，计算机不会替代外科手术，不会替代美的感觉，例如当你遥望夜空的星星时。但是，它或许会替代其他一切功能。

Can the human brain be totally free, without any form of attachment—attachment to certain beliefs, experiences and so on? If the brain cannot be totally free, it will deteriorate. When the brain is occupied with problems, with specialization, with a livelihood, it is in limited activity. But when the computer takes over, this activity will become less and less and therefore it will gradually deteriorate. This is not something in the future, it is actually happening now if one observes one's own mental activity.

人类的脑子是否能彻底自由，对于某种信仰、经验等等，没有任何形式的依附？如果脑子不能彻底自由，它就会退化。当脑子被问题、专业化和营生占据，它就处于有限的活动中了。但是如果计算机替代了它，这种活动就会变得越来越少，因而它会渐渐退化。这不是将来的事情，它现在已经发生了，如果你观察自己的心理活动的话。

>> Can your consciousness, with its basic content of fear, the pursuit of pleasure with all the implications of grief, pain and sorrow, being hurt inwardly and so on, become totally free? We may have other forms of consciousness, group consciousness, racial consciousness, national consciousness, but basically the content of our consciousness is fear, the pursuit of pleasure, with the resultant pain, sorrow and ultimate death. These comprise the central content of our consciousness. We are together observing the whole phenomenon of human existence, which is our existence. We are mankind, because our consciousness, is basically fear, the pursuit of pleasure and the never ending burden of pain, hurts, sorrow. One's consciousness is not personal to oneself. This is very difficult to accept because we have been so conditioned, so educated, that we resist the actual fact that we are not individuals at all, we are the whole of mankind. This is not a romantic idea, it is not a philosophical concept, it is absolutely not an ideal;

examined closely, it is a fact. So we have to find out whether the brain can be free from the content of its consciousness.

意识和它的基本内容——恐惧、对快乐的追逐，包括所有的挫折、痛苦、悲伤、内心受到的伤害等等，能否变得彻底自由？我们也许有其他形式的意识，群体意识、种族意识、国家意识等等，但是从根本上说，意识就是恐惧、对快乐的追逐及由此引发的痛苦、悲伤和最终的死亡。这些构成了我们意识的核心。我们是在一同观察人类生活，也就是我们生活的整个现象。我们是人类，我们的意识在根本上都负有恐惧、追求快乐和永无止境的痛苦、伤害、悲伤造成的重担。你的意识不是你个人独有的。这很难接受，因为我们是这样被约束、被教育的，因而我们抗拒真正的事实——我们根本不是个体，我们是整个人类。这不是浪漫的想法，不是哲学观念，它也完全不是一个理想，而是一个经过严密检视的事实。因此我们必须去发现，脑子是否能够从意识中解脱出来。

>> Sirs, why do you listen to the speaker? Is it that in listening to the speaker you are listening to yourself ? Is that what is taking place? The speaker is only pointing something out, acting as a mirror in which you see yourself, see the actuality of your own consciousness; it is not the description which the speaker is pointing out, which becomes merely an idea if you do no more than follow it. But if through the description, you yourself actually perceive your own state of mind, your own consciousness, then listening to the speaker has a certain importance. And if at the end of these talks you say to yourself "I have not changed; why? It is your fault. You have spoken for fifty years perhaps, and I have not changed", is it the fault of the speaker? Or you say "I have not been able to apply it; naturally it is the fault of the speaker". Then you become cynical and do all kinds of absurd things. So please bear in mind that you are listening not so much to the speaker as looking at your own consciousness through the description in

words—which is the consciousness of all humanity. The Western world may believe in certain religious symbols and certain rituals; the Eastern world does likewise, but behind it all there is the same fear, the same pursuit of pleasure, the same burden of greed, pain, of being hurt and wanting to achieve—all of which is common to the whole of humanity.

先生们，你们为什么听讲话者说这些？在听讲话者说的过程中你们在听自己吗？是正在发生什么事情吗？讲话者只是指出某些事情，他扮演了镜子的角色，因此你们可以看见自己，看见你们自己意识的真相。它不是讲话者正描述的东西，如果你们不做更多的事情而只是追随它的话，那只会变成一种观念。但是如果通过描述，你自己实际察觉到自己的心理状态、自己的意识，那么对讲话者的倾听就具有了某种重要性。而如果在这些谈话结束的时候你对自己说“为什么我没有改变？这是你的错，你说了五十来年，而我并没有改变”，这是讲话者的错吗？或者你说“我还不能运用它，这当然是讲话者的错”，那样你就会变得愤世嫉俗，并且会做各种荒唐的事情。因此请记住，与其说你在听讲话者讲话，不如说你是在通过言辞的描述去看你自己的意识——那是全人类的意识。西方社会或许信仰某种宗教符号和仪式，东方社会同样如此，但是这一切的背后有着同样的恐惧，同样的对快乐的追求，同样的贪婪、痛苦、受伤的重担，以及对成就的渴望，这一切都是整个人类共通的。

>> So, in listening we are learning about ourselves, not just following the description. We are actually learning to look at ourselves and therefore bringing about a total freedom in which the whole of the brain can operate. After all, meditation, love and compassion are the operation of the whole of the brain. When there is the operation of the whole, there is integral order. When there is integral, inward order, there is total freedom. It is only then that there can be something which is timelessly sacred. That is not a reward; that is

not something to be achieved; that which is eternally timeless, sacred, comes about only when the brain is totally free to function in wholeness.

因此，我们是在倾听中了解自己，而不只是遵循描述。我们实际上是在学习审视自己，并由此带来了整个脑子运作的完全自由。毕竟，冥想、爱和慈悲都是由整个脑子运作的。当存在整体的运作，完整的秩序就产生了。当有了完整的内在秩序，就有了完全的自由。只有这时，永恒神圣之物才能存在。这不是报偿，不是要达成的目标，只有当脑子完全自由地在整体中运作时，永恒神圣之物才会产生。

>> The content of our consciousness is put together by all the activities of thought; can that content ever be freed so that there is a totally different dimension altogether? So let us observe the whole movement of pleasure. There is not only biological, including sexual, pleasure, there is also pleasure in possessions, pleasure

in having money, pleasure in achieving something that you have been working towards; there is pleasure in power, political or religious, in power over a person; there is pleasure in the acquisition of knowledge, and in the expression of that knowledge as a professor, as a writer, as a poet; there is the gratification that comes about through leading a very strict, moral and ascetic life, the pleasure of achieving something inwardly which is not common to an ordinary man. This has been the pattern of our existence for millions of years. The brain has been conditioned to it and therefore has become limited. Anything that is conditioned must be limited and therefore the brain, when it is pursuing the many forms of pleasure, must inevitably become small, limited, narrow. And probably, unconsciously realizing this, one seeks different forms of entertainment, a release through sex, through different kinds of fulfilment. Please observe it in yourself, in your own activity in daily life. If you observe, you will see that one's brain is occupied all day with something or other, chattering, talking endlessly, going on like a machine that never stops. And in this way the brain is gradually wearing itself out—and it is going to become inactive if the computer takes its place.

我们的意识是由我们的思想活动组合在一起的，它到底有没有可能变得自由，因而产生一个完全不同的维度？让我们观察一下快乐的活动。不仅是生物学上的快乐，譬如性的快乐；还有占有的快乐、有钱的快乐、实现某种你一直努力想达到的目标的快乐；拥有政治或宗教权力的快乐、超越别人的快乐；获得知识的快乐，以及作为一个教授、作家、诗人来表达自己学问的快乐；还有通过享受一种非常严格的、道德的和苦行的生活产生的满足，获得某种不同于普通人的内心状态的快乐。这已经成了我们数百万年来的生存模式。脑子对此习以为常了，因而变得局限。任何受到约束的东西一定是受限的，因此脑子在追求各种形式的快乐时，不可避免地会变得琐碎、局限、狭隘。或许在不知不觉中意识到这一点的时候，你会通过性、通过不同的方式去寻求各种形式的娱乐，寻求释放。请在你自己身上，在你自己每天的活动中观察这件事。如果去观察，你会看到你的脑子一天到晚都被某些东西占据着，喋喋不休，没完没了地讲话，像一部永不停歇的机器那样运转。脑子就这样渐渐地耗尽了自己，如果计算机替代了它，它就会变得怠惰。

>> So, why are human beings caught in this perpetual pursuit of pleasure—why? Is it because they are so utterly lonely? Are they escaping from that sense of isolation? Is it that they have been, from childhood, conditioned to this? Is it because thought creates the image of pleasure and then pursues it? Is thought the source of pleasure? For example, one has had some kind of pleasure, eating very tasty food, or sexual pleasure, or the pleasure of being flattered and the brain registers that pleasure. The incidents which have brought about pleasure have been

recorded in the brain, and the remembrance of these incidents of yesterday, or last week, is the movement of thought. Thought is the movement of pleasure; the brain has registered incidents, pleasurable and exciting, worth remembering, and thought projects them into the future and pursues them. So the question then is: why does thought carry on the memory of an incident that is over and finished? Is not that part of our occupation? A man who wants money, power, position, is perpetually occupied with it. Perhaps, the brain is similarly occupied with the remembrance of something of a week ago which gave great pleasure, being held in the brain, which thought projects as future pleasure and pursues. The repetition of pleasure is the movement of thought and therefore limited; therefore the brain can never function wholly, it can only function partially.

那么，人类为什么会陷入这种对快乐的无尽追求之中？为什么？是因为他们非常孤独？他们是在逃避那种孤立感吗？是从童年开始他们就被约束成这样了吗？是因为思想制造了快乐的形象然后他们去追逐它吗？思想是快乐的根源吗？例如，一个人经历过某种快乐，吃过很美味的食物，或者是性的快乐，或者是被奉承的快乐，然后脑子记录了那个快乐，引起快乐的事情被脑子记录下来了，这些对昨天或上周的事情的回忆就是思想的活动。思想是快乐的活动，脑子记录了那些令人愉快的、使人兴奋的、值得记住的事情，而思想将它们投射到将来并追逐它们。那么问题是，思想为什么要携带一件已经结束和完成的事情的记忆？那不是我们占有的一部分吗？一个想要钱财、权力、地位的人会永远地被它占据。或许脑子同样被一周之前带来巨大快乐的事情占据，这些事情被装在脑子里，思想将它投射为将来的快乐去追逐。对快乐的重复是思想的活动，是受限的，因而脑子从来都不能完全地运作，它只能部分地运作。

>> Now the next question that arises is: if this is the pattern of thought, how can thought be stopped, or rather, how can the brain stop registering the incident of yesterday which gave delight? That is the obvious question, but why does one put it? Why? Is it because one wants to escape from the movement of pleasure, and that very escape is yet another form of pleasure? Whereas if you see the fact that the incident which gave great delight, pleasure, excitement, is over, that it is no longer a living thing, but something which happened a week ago—it was a living thing then but it is not so now—can you not finish with it, end it, not carry it over? It is not how to end it or how to stop it. It is just to see factually how the brain, how thought, is operating. If one is aware of that, then thought itself will come to an end. The registering of pleasure is ended, finished.

下一个出现的问题就是：如果这是思想的模式，那么要如何停止思想，更确切地说，是脑子怎么才能停止记录昨天让你高兴的事情？这是很明显的问题，但是为什么还会有人明确提出它来？为什么？是因为他想要逃避快乐的活动，而这个逃避仍然是另一种形式的快乐？然而如果你看到那个带来高兴、快乐、兴奋的事情结束了这个事实，它不再是一件当前存在的事情了，而是发生在一周之前的事情——那时它是当前的事情但现在不是了——你能不能放下它，结束它，不再延续它？不是要了解如何结束它或如何停止它，只是切实地去看思想、脑子是怎么运作的。如果你意识到这一点，那么思想自己就会终止。对快乐的记录就停止了、结束了。

>> Fear is the common state of all mankind, whether you live in a small house or in a palace, whether you have no work or plenty of work, whether you have tremendous knowledge about everything on earth or are ignorant, or whatever, there is still this deep-rooted fear which is common to all mankind. That is a common ground on which all humanity stands. There is no question

about it. It is an absolute, irrevocable fact, it cannot be contradicted. As long as the brain is caught in this pattern of fear, its operation is limited and therefore can never function wholly. So it is necessary, if humanity is to survive completely as human beings and not as machines, to find out for oneself whether it is possible to be totally free from fear.

恐惧是人类共有的状态，无论你生活在小房子里还是宫殿里，无论你是无工作还是工作繁忙，无论你对地球上的一切事情无所不知还是完全无知，都仍然会有人类共有的根深蒂固的恐惧。这是全人类的共同认知，这一点是毫无疑问的。这是一个绝对无法改变的事实，它无法被否认。只要脑子陷在这个恐惧的模式中，它的运作就是受限的，因而永远也不能完全地运作。如果人类要作为人而不是机器存活下去，那么就要亲自发现有没有可能完全从恐惧中解脱出来。

>> We are concerned with fear itself, not with the expressions of fear. What is fear? When there is fear, is there at that very moment a recognition as fear? Is that fear describable at the moment the reaction is taking place? Or does the description come after? "After" is time. Suppose one is afraid: either one is afraid of something, afraid of something that one has done in the past which one does not want another to know, or something has happened in the past which again awakens fear, or is there a fear by itself without an object? At the second when there is fear, does one call it fear? Or does that happen only afterwards? Surely it is after it has happened. Which means that previous incidents of fear which have been held in the brain are remembered immediately after the reaction takes place; the memory says "That is fear". At the immediacy of the reaction one does not call it fear. It is only after it has happened that one names it as fear. The naming of it as fear is from the remembrance of other incidents

that have arisen which have been named fear. One remembers those fears of the past and the new reaction arises which one immediately identifies with the word fear. That is simple enough. So there is always the memory operating on the present.

我们关心的是恐惧本身，不是恐惧的表现。恐惧是什么？当存在恐惧的时候，就在那个时刻，你是否能意识到恐惧？在那个反应发生的时候，那个恐惧是可描述的吗？还是说反应发生之后，你才能去描述？这个“之后”就意味着时间，假设一个人害怕，要么他害怕某件事情，害怕他过去做过的事情让另一个人知道；要么是过去发生过的事情又引起了他的恐惧。有没有一种没有对象、单独存在的恐惧？在恐惧存在的那一秒你会称它为恐惧吗？还是说是在恐惧发生之后？无疑是在它发生之后。这意味着那个储存在脑子里的以前的恐惧事件在反应发生之后被立即回忆起来了，那个记忆说“这是恐惧”。在那个即时的反应中你不会称它为恐惧，只有在它发生之后你才会将它命名为恐惧。将它命名为恐惧是因为你记得其他发生过的被称为恐惧的事件。你想起了过去的恐惧，于是新的反应发生了，你立即将它和“恐惧”这个词联系起来。这非常简单。因此过去总是对现在产生着作用。

>> So is fear time?—the fear of something which happened a week ago, which has caused that feeling which we have named as fear and the future implication that it must not happen again; yet it might happen again, therefore one is afraid of it. So one asks oneself: is it time that is the root of fear?

那么，恐惧是时间吗？对一周以前发生的事情的恐惧导致了那种我们称之为恐惧的感觉，并牵涉到将来——它不能再发生了，然而它或许还会发生，所以你害怕它。因此我问自己：时间是恐惧的根源吗？

>> So what is time? Time by the watch is very simple. The sun rises at a certain time and sets at a certain

time—yesterday, today and tomorrow. That is a natural sequence of time. There is also psychological, inward time. The incident which happened last week, which has given pleasure, or which awakened the sense of fear, is remembered and projected into the future—I may lose my position, I may lose my money, I may lose my wife—time. So is fear part of psychological time? It looks like it. And what is psychological time? Not only does physical time need space, but psychological time needs space as well—yesterday, last week, modified today, tomorrow. There is space and time. That is simple. So, is fear the movement of time? And is not the movement of time, psychologically, the movement of thought? So thought is time and time is fear—obviously. One has had pain sitting with the dentist. It is stored, remembered, projected; one hopes not to have that pain again—thought is moving. So fear is a movement of thought in space and time. If one sees that, not as an idea, but as an actuality (which means one has to give to that fear complete attention at the moment it arises) then it is not registered. Do this and you will find out for yourself. When you give complete attention to an insult, there is no insult. Or if somebody comes along and says, "What a marvellous person you are" and you pay attention it is like water off a duck's back. The movement of fear is thought in time and space. That is a fact. It is not something described by the speaker. If you have observed it for yourself, then it is an absolute fact, you cannot escape from it. You cannot escape from a fact, it is always there. You may try to avoid it, you may try to suppress it, try every kind of escape, but it is always there. If you give complete attention to the fact that fear is the movement of thought, then fear is not, psychologically. The content of our consciousness is the movement of thought in time and space. Whether that thought is very limited, or wide and extensive, it is still a movement in time and space.

那么什么是时间？钟表的时间是非常简单的。太阳在某个时间升起，又在某个时间落下——昨天、今天和明天都如此。那是一个自然的时间顺序。还有心理上的时间。上周发生的引起快乐或恐惧的事件被回忆起来并投射到将来——我也许会失去我的职位，失去我的钱，失去我的妻子。因此恐惧是心理时间的一部分吗？看上去它是。什么是心理时间？不仅物理时间需要空间，心理时间也需要空间——昨天、上周、修饰过的今天、明天。有时间就有空间，这很简单。那么，恐惧是时间的活动吗？心理时间的活动不就是思想的活动吗？因此思想就是时间，而时间就是恐惧，很显然。你曾经在牙医那里遭受疼痛，它被存储、记住、投射。你希望不再经历那个疼痛——这是思想在活动。因此恐惧是思想在空间和时间里的运动。如果你把它看作一个事实而不是一个想法（这意味着在恐惧升起的时刻，你必须给予它完全的注意），那么它就不会被记录。这样做，你自己会找到答案。当你对一个损害给予完全的注意，那么就没有损害。或者如果某人走过来说“你是一个多么了不起的人啊”，你注意了，那么它就会成为一阵耳旁风。恐惧的活动是时间和空间中的思想，这是事实，不是讲话者的某种形容。如果你亲自观察它，那么它就是一个绝对的事实，你无法回避它。你无法回避一个事实，它总在那里。你可以尝试避开它、压制它，尝试每一种逃避，但是它始终在那里。如果你对“恐惧是思想的活动”这个事实给予完全的注意，那么恐惧就在心理上消失了。我们的意识是思想在时间和空间中的活动。无论那个思想非常有限还是非常广阔，它仍然是一种时空中的活动。

>> Thought has created many different forms of power in ourselves, psychologically, but they are all limited. When there is freedom from limitation, there is an astonishing sense of power, not mechanical power but a tremendous sense of energy. It has nothing to do with thought and therefore that power, that energy

cannot be misused. But if thought says, "I will use it", then that power, that energy, is dissipated.

思想在我们心理上制造了多种形式的力量，但是它们都是受限的。当从限制中解脱出来，会产生一种令人吃惊的力量感，不是机械的力量，而是一种惊人的能量感。它和思想没有任何关系，因而那种力量、那种能量不可能被误用。但是如果思想说"我要运用它"，那么那种力量、那种能量就消失了。

>> Another factor which exists in our consciousness is sorrow, grief, pain and the wounds and hurts that remain in most human beings from childhood. That psychological hurt, the pain of it, is remembered, it is held on to; grief arises from it; sorrow is involved in it. There is the global sorrow of mankind which has faced thousands and thousands of wars, for which millions of people have cried. The war machine is still with us, reinforced by our nationalism, by our feeling that we

are separate from the rest, "we" and "them", "you" and "me". We are ready for another war and when we prepare for something, there must be some kind of explosion somewhere—it may not be in the Middle East, it may happen here. As long as we are preparing for something, we are going to get it—it is like preparing food. But we are so stupid that all this goes on—including terrorism.

存在于我们意识中的其他因素是悲伤、痛苦，以及从童年时期就保留在大多数人身上的创伤和损害。那种心理上的伤害及它的痛苦，被记住了、被保留了，由此产生了痛苦，也牵涉到悲伤。全球人类都存在悲伤，即人类经历了成千上万次战争，无数人因此而哭泣。战争机器仍然伴随着我们，被我们的国家主义强化，被我们同其他部分的分裂感——“我们”和“他们”、“你”和“我”的分裂感——所强化。我们在准备着另一场战争，当我们在准备的时候，某个地区必定存在扩张——它或许不是在中东，而是发生在这里。只要我们在准备什么东西我们就会得到它，就像准备食物一样。但是我们是如此愚蠢，以至于这一切都在发生，包括恐怖主义。

>> We are asking whether this whole pattern of being hurt, knowing loneliness and pain, resisting, withdrawing, isolating ourselves, which causes further pain, can come to an end; whether the grief, the sorrow of losing some precious belief that we have held, or the disillusionment that comes when we lose somebody we have followed, for whom we have struggled, surrendered ourselves, can also come to an end? Is it possible ever to be free of all this? It is possible if we apply ourselves, not just endlessly talk about it. As it is we realize that we are hurt psychologically from childhood, we see all the consequences of that hurt, which we resist, from which we withdraw, not wanting to be hurt any more. We encourage isolation and therefore build a wall round ourselves. In our relationships we are doing the same thing.

我们要问，知道了孤独、痛苦、对抗、退缩、自我隔绝——这会导致进一步的痛苦，那这受伤害的整个模式能不能停止？还有，对于失去曾经拥有的宝贵信仰的悲伤惋惜，或者因失去我们追随的人而产生幻灭——我们曾经为他奋斗、放弃自己，这些能够停止吗？到底有没有可能从这一切中解脱？如果我们对自己下功夫，而不只是没完没了地谈论它，这就是可能的。我们其实了解自己从童年时期就受到的心理上的伤害，我们能看到那些伤害的一切后果，我们抗拒它，我们从中退出，因为我们不想再受伤害。我们鼓励隔绝，于是围绕自己建起了一堵墙。在人际关系中我们也在做着同样的事情。

>> The consequences of being hurt from childhood are pain, resistance, withdrawal, isolation, deeper and deeper fear. And, as the speaker has said, there is the global sorrow of mankind; human beings have been tortured through wars, tortured under dictatorships, totalitarianism, tortured in different parts of the world.

And there is the sorrow of my brother, son, wife, running away, or dying; the sorrow of separation, the sorrow that comes about when one is deeply interested in something and the other is not. In all this sorrow there is no compassion, there is no love. The ending of sorrow brings love—not pleasure, not desire, but love. Where there is love, there is compassion with which comes intelligence, which has nothing whatever to do with the "intelligence" of thought.

童年伤害的后果就是痛苦、抗拒、逃避、隔绝，以及不断加深的恐惧。就像讲话者说过的，这是人类整体的悲伤。人类被战争折磨、被独裁和专制折磨，在这个世界上的各个地区受到折磨。悲伤无处不在：我兄弟、儿子、妻子的悲伤，逃跑或死亡、分离的悲伤，当你对一件事情兴趣浓厚而别人却不以为然的时候产生的悲伤。所有的悲伤中没有同情，也没有爱，悲伤的终结会带来爱，不是带来快乐、欲望，而是爱。有爱的地方就有慈悲，智慧也会随之而来，它和思想的“智慧”毫不相干。

>> We have to look very closely at ourselves as humanity, at why we have borne all these things all our lives, at why we have never ended this condition. Is it part indolence, part habit? We generally say: "It is part of our habit, part of our conditioning. What am I to do about it? How am I to uncondition myself? I cannot find the answer; I will go to the guru next door."—or further away, or the priest, or this or that. We never say: "Let us look at ourselves closely and see if we can break through it, like any other habit." The habit of smoking can be broken, or that of drugs and alcohol. But we say: "What does it matter? I am getting old anyhow, the body is destroying itself, so what does a little more pleasure matter?" So do we carry on. We do not feel utterly responsible for all the things we do. We either blame it on the environment, on society, on our parents, on past hereditary; we find some excuse but never apply ourselves. If we really have the urge, the immediate urge, to find out why we are hurt, it can be done. We

are hurt because we have built an image about ourselves. That is a fact. When one says, "I am hurt", it is the image that one has about oneself that is hurt. Somebody comes along and puts his heavy boot on that image and one gets hurt. One gets hurt through comparison: "I am this, but somebody else is better." As long as one has an image about oneself, one is going to get hurt. That is a fact and if one does not pay attention to that fact, but retains an image of oneself of any kind somebody is going to put a pin into it and one is going to get hurt. If one has an image about oneself as addressing large audiences and being famous, having gained a reputation which one wants to maintain, then someone is going to hurt it—somebody else with a bigger audience. If one gives complete attention to the image, one has about oneself—attention, not concentration but attention—then one will see that the image has no meaning and it disappears.

21st July, 1981

作为人类，我们必须密切审视自己，看看为什么我们承担了整个生命中的这一切，为什么我们从未终止这种约束。一部分是因为懒惰，一部分是因为习惯吗？通常我们会说“这是我们习惯的一部分，约束的一部分。我要怎么办呢？我要如何解除自己的约束？我找不到答案，我要去找隔壁的古鲁”，要么就越走越远，要么就去找牧师，或其他什么人。我们从来不说“让我们密切地审视自己，看看我们能不能突破它，就像对别的习惯一样”。抽烟的习惯能够被打破，吸毒和酗酒也能。但是我们说：“这有什么关系？我已经老了，身体渐渐垮了，所以多一点快乐又有什么关系呢？”所以我们继续原来的样子。我们完全没觉得要对自己所做的一切负责。我们还将它归咎于环境、社会、我们的父母、过去的遗传。我们总是找一些借口，却从来不对自己下功夫。如果我们真的拥有急切的欲望，去探明我们为什么受伤，它就能够被解决。我们受伤是因为我们建立了关于自身的形象。这是一个事实。当我们说“我受伤了”，其实是那个你拥有的关于自己的形象受伤了。当某人把他沉重的皮靴踩在那个形象上面，你就受伤了。你是因为比较而受伤的，“我是这样的，而别人更好”。只要想着关于自身的形象，你就会受伤。这是一个事实，如果你没有注意到这个事实，而是保留了一个任何人都可以把针扎进去的关于自身的形象，那么你就会受伤。如果你建立的关于自己的形象是一个拥有大批粉丝的名人，获得了自己想要维护的名誉，那么有人就会伤害它——那些拥有更多粉丝的人。如果你把全部的注意放在这个关于自己的形象上——是注意，不是专注，那么你就会看到这个形象是没有意义的，然后它就消失了。

一九八一年七月二十一日

>> I think we ought to talk over together, going into it rather deeply, the implication of sorrow, so as to find out for ourselves whether sorrow and love can exist together. And also what is our relationship to the sorrow

of mankind?—not only to our own personal daily grief, hurt, pain, and the sorrow that comes with death. Mankind has suffered thousands of wars; there seems to be no end to wars. We have left it to the politicians, all over the world, to bring about peace, but what they are doing, if you have understood them, will never bring peace. We human beings have never been able to live in peace with each other. We talk about it a great deal. The religions have preached peace—peace on earth and goodwill—but apparently it has never been possible to have peace on earth, on the earth on which we live, which is not the British earth or the French earth, it is our earth. We have never been able to resolve the problem of killing each other.

为了探明悲伤和爱是否能够同时存在，我想我们应该一同来更加深入地探究悲伤的含义，以及我们和悲伤的关系是什么，不只是我们和自己个人的日常生活中的伤害、痛苦以及伴随死亡而来的悲伤的关系。人类承受了成千上万次战争，战争似乎永无止境。我们希望世界上的政治家能带来和平，但是如果你了解他们的话，就会发现他们的所作所为永远也不会带来和平。我们人类从来没能和平共处。我们总在谈论和平。宗教宣扬和平——全世界的和平与善意，但是，在这个世界上，在我们生活的地球上——不是英国人的地球或法国人的地球，而是我们的地球——和平显然从未实现。我们从来没能解决互相杀戮的问题。

>> Probably we have violence in our hearts. We have never been free from a sense of antagonism, a sense of retaliation, never free from our fears, sorrows, wounds and the pain of daily existence; we never have peace and comfort, we are always in travail. That is a part of our life, of our daily suffering. Man has tried many many ways to be free of this suffering without love; he has suppressed it, escaped from it, identified himself with something greater, handed himself over to some ideal,

or belief or faith. Apparently this sorrow can never end; we have become accustomed to it, we put up with it, we tolerate it and we never ask ourselves seriously, with a great sense of awareness, whether it is possible to end it.

或许我们内心隐藏着暴力。我们从来没有摆脱敌对、报复的感觉，从来没有摆脱恐惧、悲痛、创伤和日常生活中的痛苦，从未拥有和平与舒适，我们总是在受苦。那是我们生命的一部分，我们日常生活的一部分。人类尝试了很多很多年，要从这种没有爱的情境中解脱出来；我们压制它、逃避它，将自己与某个更大的东西联系起来，把自己交给某个理想、信念或信仰。但悲伤显然从未终止。我们变得对它习以为常，我们容忍它，却从来不带着充分的觉察认真地问自己，有没有可能结束它。

>> We should also talk over together the immense implications of death. Death is a part of life, though we

generally postpone or avoid even talking about it. It is there and we ought to go into it. And we should also enquire whether love—not the remembrance of pleasure which has nothing to do with love and compassion—whether love with its own peculiar all-comprehending intelligence can exist in our life.

我们还应该一同讨论死亡的深刻含义。死亡是生命的一部分，尽管我们通常会延缓，甚至避免去谈论它。但它就在那儿，我们应当探究它。我们也应该探寻爱，不是探寻快乐的记忆，那和爱、慈悲毫不相关，而是探寻爱以及领悟它特有的整体性的智慧有没有可能在我们的生活中出现。

>> First of all, do we, as human beings, want to be really free from sorrow? Have we ever actually gone into it, faced it and understood all the movement of it, the implications involved in it? Why is it that we human beings—who are so extraordinarily clever in the

technological world—have never resolved the problem of sorrow? It is important to talk this question over together, and find out for ourselves whether sorrow can really end.

首先，作为人类，我们真的能从悲伤中解脱出来吗？我们真正探究过、面对过、了解过它所有的活动，以及其中牵涉的含义吗？人类在技术世界出奇地聪明，但为什么从未解决悲伤的问题？一起讨论这个问题，并亲自发现悲伤能否真正终止，这是非常重要的。

>> We all suffer in various ways. There is the sorrow for death of someone, there is the sorrow of great poverty, and the great sorrow of ignorance—“ignorance” not in the sense of book knowledge but the ignorance of not knowing oneself totally, the whole complex activity of the self. If we do not understand that very deeply, then there remains the sorrow of that ignorance. There

is the sorrow of never being able to realize something fundamentally, deeply—though we are very clever at achieving technological success and other successes in this world. We have never been able to understand pain, not only physical pain, but the deep psychological pain, however learned or not very erudite we may be. There is the sorrow of constant struggle, the conflict from the moment we are born until we die. There is the personal sorrow of not being beautiful outwardly or inwardly. There is the sorrow of attachment with its fear, with its corruption. There is the sorrow of not being loved and craving to be loved. There is the sorrow of never realizing something beyond thought, something which is eternal. And ultimately there is the sorrow of death.

我们都在以各种不同的方式受苦。某人死去导致的悲伤，严重贫困带来的悲伤，无知带来的巨大悲伤——不是书本知识意义上的“无知”，而是不完全了解自己、自我的整个复杂活动的无知。如果我们不是非常深入地了解自己，那么无知就会带来悲伤。虽然我们在技术和其他方面取得了成功，但我们永远无法从根本上深刻认识到一些事情，这是一种悲哀。我们从未了解痛苦，不只是身体上的痛苦，还有深刻的精神痛苦，无论我们博学还是孤陋寡闻。还有不断争斗带来的悲伤，从生到死一直都存在的冲突带来的悲伤，外在或内在不美丽带来的个人化的悲伤，还有依附于它带来的恐惧、腐朽导致的悲伤，不被爱而渴求被爱的悲伤，从未认识到某种超越思想的、永恒的事物的悲伤，以及最终对于死亡的悲伤。

>> We have described various forms of sorrow. The basic factor of sorrow is self-centred activity. We are all so concerned with ourselves, with our endless problems, with old age, with not being able to have a deep inward yet global outlook. We all have images of ourselves and of others. The brain is always active in daydreaming, being occupied with something or other, or creating

pictures and ideas from the imagination. From childhood one gradually builds the structure of the image which is “me”. Each one of us is doing this constantly; it is that image, which is “me”, that gets hurt. When the “me” is hurt, there is resistance, the building of a wall round oneself so as not to be hurt any more; and this creates more fear and isolation, the feeling of having no relationship, the encouraging of loneliness which also brings about sorrow.

我们描述了各种形式的悲伤。悲伤是以自我为中心的活动。我们全都关心自己和自己那些没完没了的问题，衰老、缺乏深刻的内在和全局性的视野。我们都有关于自己和他人的形象。脑子总是在幻想中忙碌着，被一些事物占据，或者根据想象来建立画面和观念。一个人从童年开始逐步建立了“我”的形象结构。我们每个人都在持续不断地做这件事。正是那个形象，也就是“我”受到了伤害。“我”受伤的时候会有抗拒，为了不再受伤而在周围建起一堵墙，而这导致了进一步的恐惧和隔绝，没有人际链接的感觉，以及对孤独的鼓励，这也会带来悲伤。

>> After having described the various forms of sorrow, can we look at it without verbalization, without running away from it into intellectual adaptation to some form of religious or intellectual conclusion? Can we look at it completely, not moving away from it, but staying with it? Suppose I have a son who is deaf or blind; I am responsible, and it gives sorrow knowing that he can never look at the beautiful sky, never hear the running waters. There is this sorrow: remain with it, do not move away from it. Or suppose I have great sorrow for the death of someone with whom I have lived for many years. Then there is this sorrow which is the essence of isolation; we feel totally isolated, completely alone. Now, remain completely with that feeling, not verbalizing it, not rationalizing it, or escaping from it, or trying to transcend it—all of which is the

movement that thought brings about. When there is that sorrow and thought does not enter into it at all—which means that you are completely sorrow, not trying to overcome sorrow, but totally sorrow—then there is the disappearance of it. It is only when there is the fragmentation of thought that there is travail.

在描述了各种形式的悲伤之后，我们能不能不再用语言来描述它，不再因逃避它而在智力上陷入某种形式的宗教或智力的结论中？我们能不能完整地看一看它，不要离开，而是和它共处？假设我有个儿子，他是聋哑人或盲人，我尽职尽责，而他永远无法看到美丽的天空，永远无法听到流水的潺潺声，这会带来悲伤，和这种悲伤在一起，不要离开它。或者我因为某个多年一起生活的人死去而产生强烈的悲伤。然后就有了这种作为孤独本质的悲伤，我们感到完全、彻底的孤独。现在，完全和那个感觉在一起，不去描述它，不把它合理化，或者逃离它，或者试图超越它——所有那些都是思想引起的活动。

当悲伤在那里，而思想完全没有参与进来的时候——这意味着你是全然的悲伤，不要试图压制悲伤，而是尽情去悲伤——那时悲伤就会消失。只有思想碎片存在的时候，痛苦才会存在。

>> When there is sorrow, remain with it without a single movement of thought so that there is the wholeness of it. The wholeness of sorrow is not that I am in sorrow, I am sorrow—and then there is no fragmentation involved in it. When there is that totality of sorrow, no movement away from it, then there is the withering away of it.

与悲伤共处，不要有任何思想的活动，这样悲伤就是整体的。整体的悲伤并不是我处于悲伤之中，而是我就是悲伤，那样就不存在分裂。当存在完整的悲伤而没有逃离它的活动，悲伤就会消失。

>> Without ending sorrow how can there be love? Strangely we have associated sorrow and love. I love my son and when he dies, I am full of sorrow—sorrow we associate with love. Now we are asking: when there is suffering, can love exist at all? But is love desire? Is love pleasure—so that when that desire, that pleasure, is denied, there is suffering? We say that suffering as jealousy, attachment, possession, is all part of love. That is our conditioning, that is how we are educated, that is a part of our inheritance, tradition. Now, love and sorrow cannot possibly go together. That is not a dogmatic statement, or a rhetorical assertion. When one looks into the depth of sorrow and understands the movement of it in which is involved pleasure, desire, attachment, and the consequences of that attachment, which bring about corruption, when one is aware without any choice, without any movement, aware of the whole nature of sorrow, then can love exist with sorrow? Or is love something entirely different? We ought to be clear that

devotion to a person, to a symbol, to the family, is not love. If I am devoted to you for various reasons, there is a motive, behind that devotion. Love has no motive. If there is a motive it is not love, obviously. If you give me pleasure, sexually, or various forms of comfort, then there is dependency; the motive is my dependence on you because you give me something in return; and as we live together I call that love. Is it? So one questions the whole thing and asks oneself: where there is motive can love exist?

没有悲伤的终结，爱怎么能存在？不可思议的是，我们将悲伤和爱联系在了一起。我爱我的儿子，当他死去的时候我充满了悲伤——我们将它和爱联系起来了。我们现在要问：当痛苦存在的时候，爱能够存在吗？爱是欲望吗？爱是快乐吗？这样在欲望、快乐被否定的时候会产生痛苦吗？我们说嫉妒、依附、占有，这些都是爱的表现。那是我们的约束，我们就是那样被教育的，那是我们遗传的、传统的一部分。因此，爱和悲伤不可能相伴。这不是一个武断的陈述，或一个夸张的定论。当你深入地检视悲伤，了解它的活动，其中牵涉快乐、欲望、依附以及依附带来的后果，它导致腐朽，当你没有任何选择，没有任何活动，觉察到悲伤的整个本质的时候，爱还会裹挟着悲伤而存在吗？还是说爱是某种完全不同的东西？我们应该清楚，专注于人、符号，专注于家庭，那都不是爱。如果我由于各种原因投入精力在你身上，这种投入的背后就有一个动机。爱没有动机，如果有动机，那么这很显然就不是爱。如果你给我快乐、性或各种形式的舒适，那么就有了从属；动机是我对你的依赖，因为你给我某种回报，因为我们生活在一起，我称之为爱。但它是吗？于是你质疑整件事情，并问自己：动机存在的地方，爱能存在吗？

>> Where there is ambition, whether in the physical world, or in the psychological world—ambition to be on top of everything, to be a great success, to have power,

religiously, or physically—can love exist? Obviously not. We recognize that it cannot exist and yet we go on. Look what happens to the brain when we play such tricks. I want to achieve illumination—you know, you cannot achieve illumination; you cannot possibly achieve that which is beyond time. Competitiveness, conformity, jealousy, fearfulness, hate, all that is going on, psychologically, inwardly. We are either conscious of it, or we deliberately avoid it. Yet I say to my wife or father, whoever it is, "I love you". What happens when there is such deep contradiction in my life, in my relationship? How can that contradiction have any sense of deep integrity? And yet that is what we are doing until we die. Can one live in this world without ambition, without competitiveness? Look at what is happening in the outward world. There is competition between various nations; the politicians are competing with each other, economically, technologically, in building up the instruments of war; and so we are destroying ourselves. We allow this to go on because we are also inwardly competitive.

在有野心的地方，无论是物质世界还是心理世界，想要超凡脱俗、取得巨大成功、拥有权力，爱能够存在吗？显然不能。我们承认它不能存在，却依然追寻它。看看当我们玩这个把戏的时候脑子里发生了什么。我想获得觉悟——你知道的，你不可能获得觉悟，你不可能获得那个超越时间的东西。竞争、服从、嫉妒、恐惧、仇恨，在心理上，那一切都在持续。我们要么意识到它，要么故意避开它。然而我对我的妻子或父亲或无论什么人说“我爱你”。在我的生活中，在我们的关系中，出现这样深刻的矛盾时，会发生什么呢？这矛盾怎么能具有深刻的完整性呢？那就是我们一直到死都在做的事情。一个人能不能没有野心、没有竞争地活在这个世界上？看看外部世界都在发生着什么。不同国家间的竞争，政客之间相互抗争，在经济上、技术上、建造战争机器上，因此我们是在毁灭自己。我们之所以允许这些继续，是因为我们内心也是竞争性的。

>> As we pointed out, if a few really understand what we have been talking about for the last fifty years, and are really deeply involved and have brought about the end of fear, sorrow and so on, then that will affect the whole of the consciousness of mankind. Perhaps you are doubtful whether it will affect the consciousness of mankind? Hitler and his kind have affected the consciousness of mankind—Napoleon, the Caesars, the butchers of the world have affected mankind. Also the good people have affected mankind—I do not mean respectable people. The good are those who live life wholly, not fragmented. The great teachers of the world have affected human consciousness. But if there was a group of people who had understood what we have been talking about—not verbally but actually living life with great integrity—then it would affect the whole consciousness of man. This is not a theory.

This is an actual fact. If you understand that simple fact you will see that it goes right through; television, newspapers, everything, is affecting the consciousness of man. So love cannot exist where there is a motive, where there is attachment, where there is ambition and competitiveness, love is not desire and pleasure. Just feel that, see it.

我们曾经指出过，如果有几个人真的理解过去五十年我们谈论的东西，并且真正进行深入探索，实现了恐惧、悲伤等的终结，那么他们将影响人类的整体意识。或许你对此深表怀疑，对吗？希特勒和他的同僚影响了人类的意识，拿破仑、恺撒这些刽子手影响了人类。好人也影响了人类，我不是指受人尊敬的人，是指那些过着完整的而不是破碎的生活的人。世界上伟大的导师影响了人类的意识。但是如果有一群人理解了我们所说的——不是字面上理解，而是真正带着巨大的诚意去体验生活，那么他们就会影响人类的整体意识。这不是理论，这是一件实际的事情。

如果你了解那个简单的事实，你就会看到，它是真实存在而不虚幻的。电视、报纸，每件事情都在影响人类的意识。在有动机、依附、野心和竞争的地方，爱不可能存在，爱不是欲望和快乐。所以，去感受它、去看它就可以了。

>> We are going into all this so as to bring about order in our life—order in our "house", which has no order. There is so much disorder in our life and without establishing an order that is whole, integral, meditation has no meaning whatsoever. If one's "house" is not in order, one may sit in meditation, hoping that through that meditation one will bring about order; but what happens when one is living in disorder and one meditates? One has fanciful dreams, illusions and all kinds of nonsensical results. But a sane, intelligent, logical man, must first establish order in daily life, then he can go into the depths of meditation, into the meaning and the beauty of it, the greatness of it, the worth of it.

探究所有这些是为了在我们的生活中带来秩序——我们“房子”里的秩序，不过那里本没有秩序。我们生活中有那么多的混乱，不建立一个完全的、整体的秩序，冥想就没有任何意义。如果你的“房子”里没有秩序，你也许会坐下来冥想，希望通过冥想带来秩序。但是当你在混乱的状态下冥想，会发生什么呢？你会得到奇幻的梦、错觉和各式各样荒谬的结果。但是一个心智健全的、理智的、合乎逻辑的人，一定会首先在日常生活中建立秩序，然后他才能够探究冥想的深度，它的意义，它的美，它的浩瀚和它的价值。

>> Whether we are very young, middle-aged or old, death is part of our life, just as love, pain, suspicion, arrogance, are all part of life. But we do not see death as part of our life; we want to postpone it, or put it as far away from us as possible, so we have a time interval between life and death. What is death? This question is again rather complex.

无论我们处于青年、中年还是老年时期，死亡都是我们生活中的一部分，就像爱、痛苦、怀疑、傲慢是我们生活中的一部分一样。但是我们没有把死亡看作生活中的一部分，我们想要推迟它，或者尽可能让它远离我们，因此在生活和死亡之间，我们有了一个时间间隔。什么是死亡？这又是一个很复杂的问题。

>> The Christian concept of death and suffering and the Asiatic conclusion about reincarnation are just beliefs and like all beliefs they have no substance. So put those aside and let us go into it together. It may be unpleasant; you may not want to face it. You are living now, healthily, having pleasure, fear, anxiety and tomorrow there is hope and you do not want to be concerned with the ending of all this. But if we are intelligent, sane, rational, we have to face not only the living and all the implications of the living, but also the implications of dying. We must know both. That is the wholeness of life in which there is no division. So what

is death apart from the physical ending of an organism that has lived wrongly, addicted to drink, to drugs and overindulgence or asceticism and denial? The body goes through this constant battle between the opposites, it has not experienced a balanced harmonious life, but one of extremes. Also the body goes through great stress imposed by thought. Thought dictates and the body is controlled thereby; and thought being limited brings about disharmony. It causes us to live in disharmony physically, forcing, controlling, subjugating, driving the body—this is what we are all doing including fasting for political or religious reasons, which is violence. The body may endure all this for many years, reaching old age and not getting senile. But the body will inevitably come to an end, the organism will die; is that what death is? Is the coming to an end of the organism, either through some disease, old age or accident, what we are concerned about? Is it that thought identifies itself with the body, with the name, with the form, with all the memories, and says, "Death must be avoided"? Is it that we are afraid of the coming to an end of a body that has been looked after, cared for? Perhaps we are not afraid of that especially, perhaps slyly anxious about it, but that is not of great importance. What is far more important for us is the ending of the relationships that we have had, the pleasures that we have had, the memories, pleasant and unpleasant, all of which make up what we call living—the daily living, going to the office, the factory, doing some skilful job, having a family, being attached to the family, with all the memories of that family, my son, my daughter, my wife, my husband, in the family unit. There is the feeling of being related to somebody, though in that relationship there may be great pain and anxiety; the feeling of being at home with somebody; or not at home with anybody. Is that what we are afraid of?—the ending of my relationships, my attachments, the ending of something I have known, something to which I have clung, something in which I have specialized all my life—am I afraid of the ending of all that? That is the ending of all that is "me"—the family, the name, the home, the tradition, the inheritance, the cultural education and racial inheritance, all that is "me", the "me" that is struggling or that is happy. Is that what we are afraid of?—the ending of "me", which is the ending, psychologically, of the life which I am leading, the life which I know with its pain and sorrow. Is that what we are afraid of?

关于死亡和受苦的基督教观念，以及关于转世的东方观念，这些只是信念。和所有的信念一样，它们并不基于事实。所以让我们将它们抛在一边，共同来探究。这也许使人不快，你可能不愿面对它。你现在活得很健康，有自己的快乐、恐惧、忧虑，以及对明天的希望，你不想去关心如何让这一切停止。但是如果我们是聪明的、理智的、清醒的，我们就必须去面对，不仅要面对生活以及它的全部内涵，也要面对死亡。我们必须两者都认识。那才是整体的生活，其中没有分隔。因为过着不健康的生活，沉溺于酒精、毒品、自我放纵或禁欲克制，导致有机体的生理性终结，除此之外，死亡是什么呢？身体在对立当中经受了这些持续的挣扎，没有经受过平稳和谐的生活，而总是在两个极端之间徘徊。身体还承受了思想施加的巨大压力。思想支配，身体受控，而局限的思想导致了不和谐。它使我们活在生理失调当中，强迫、控制、征服、驱使身体，这是我们所有人都在做的，包括为了政治或宗教的原因而禁食，这是暴力。身体也许会忍受这一切很多年，一直到年老而不衰竭。但身体不可避免地会终结，机体会死亡。这就是对什么是死亡的回答吗？因为疾病、年老或事故导致有机体的终结？我们关心的是什么？是思想认同身体、名字、形式、所有的记忆，并且说“必须避免死亡”吗？是我们害怕那个需要照顾和关注的身体的终结吗？或许我们并不特别害怕它，或许我们偷偷地渴望它，但是那没有多么重要。对我们来说更为重要的是，我们曾经拥有的关系、快乐以及愉快的和不愉快的记忆的结束，那一切构成了被我们称为生活的东西——日常生活，去办公室、去工厂；从事熟练的工作，拥有一个家庭；从属于那个家庭和它所有的记忆，在家庭中，有我的儿子、我的女儿、我的妻子或丈夫。那里还有与某人相互关联的感觉，尽管在那个关系中可能有着巨大的痛苦和忧虑，和某人一起无拘无束的感觉，或者和某人一起不是无拘无束的感觉。那是我们害怕的东西吗？——我的关系、我的依附的终止，我知道的东西、我执着的东西、我终生所从事的东西的终止——我害怕那一切的终止吗？那是“我”的一切的终止——家人、名字、家庭、传统、遗产、文化教育和种族遗传，那一切都是“我”，那个或挣扎或快乐的“我”。我们害怕的是它——“我”的终结，心理上就是我正在过的生活的终结，我知道的痛苦和悲伤的生活的终结。我们害怕的是它吗？

>> If we are afraid of that and have not resolved that fear, still death inevitably comes, then what happens to that consciousness, which is not your consciousness but the consciousness of mankind, the consciousness of the vast

whole of humanity? As long as I am afraid as an individual with my limited consciousness, it is that that I am afraid of. It is that of which I am scared. One realizes that it is not a fact that one's consciousness is totally separate from that of everybody else—one sees that separateness is an illusion, it is illogical, unhealthy. So one realizes, perhaps in one's heart, in one's feeling, that one is the whole of mankind—not an individual consciousness, which has no meaning. And one has lived this kind of life, which is pain, sorrow, anxiety, and if one's brain has not transformed some of all that, one's life is only a further confusion to the wholeness. But if one realizes that one's consciousness is the consciousness of mankind, and that for the human consciousness one is totally responsible, then freedom from the limitation of that consciousness becomes extraordinarily important. When there is that freedom, then one is contributing to the breaking down of the limitation of that consciousness. Then death has a totally different meaning.

23rd July, 1981

如果我们害怕它而没有解决那个恐惧，直到死亡不可避免地到来，那时候意识会发生什么？它不是你的意识而是人类的意识，是人类浩瀚整体的意识。只要我作为个体带着我有限的意识而感到恐惧，那就是我所害怕的。把我吓倒的就是它。你认识到“你的意识和其他人的意识是完全分离的”这个说法并不是事实，你看到那个分离是一个错觉，是不合逻辑的、不正常的。也许在你的心里，在你的感受里，此刻你会认识到，你就是整个人类，因为个别的意识没有意义。而你经历了这样的生活，痛苦、悲伤、忧虑，如果你的脑子没有改变其中的一些东西，那么你的生活对整体来说就只是更进一步的混乱。但是如果你认识到，你的意识就是人类的意识，并且你对人类的意识负有完全的责任，那么从那个意识的局限中获得自由就变得极其重要了。这份自由存在的时候，你就是在为消除那个意识的局限贡献力量。那时死亡就有了一种完全不同的意义。

一九八一年七月二十三日

>> One has lived a so-called individual life, concerned about oneself and one's problems. Those problems never end, they increase. One has lived that kind of life. One has been brought up, educated, conditioned, to that kind of life. You come along as a friend—you like me, or you love me—you say to me: "Look, your consciousness is not yours; you suffer as other people suffer." I listen to it and I do not reject what you say, for it makes sense, it is sane and I see that in what you have told me there can perhaps be peace in the world. And I say to myself: "Now, can I be free from fear?" I see that I am responsible, totally, for the whole of consciousness. I see that when I am investigating fear, I am helping the total human consciousness to lessen fear. Then death has a totally different meaning. I am living a life which is not my particular life. I am living a life of the whole of humanity and if I understand death, if I understand grief, I am cleansing the whole of the consciousness of mankind. That is why it is important to understand the

meaning of death and perhaps to find that death has great significance, great relationship with love, because where you end something love is. When you end attachment completely then love is.

23rd July, 1981

我过着一种所谓的个人生活，关心自己和自身的问题。那些问题从未结束，它们在增加。我过着那种生活。我被养大，受到教育，被制约，过着那样的生活。你作为一个朋友出现了，你喜欢我，或者爱我，你对我说："看，你的意识不是你的，你受苦，别人也会受苦。"我听到它，并且不排斥你说的，因为它有意义，它是理智的，并且在你的话中我看到或许可能存在世界和平。我对自己说："现在，我能否从恐惧中解脱？"我看到我对整体意识负有完全的责任。我看到，当我研究恐惧的时候，我是在帮助整个人类意识减少恐惧。那时死亡就有了一种完全不同的意义。我过的不是我的特别的生活。我过着人类整体的生活，如果我了解死亡，如果我了解悲伤，那么我就是在净化人类的整体意识。那就是为什么了解死亡的意义是重要的，弄清楚死亡是什么具有重大的意义，它和爱有着巨大的关系，因为在你死亡的地方爱就出现了。当你彻底结束依附状态的时候，爱就出现了。

一九八一年七月二十三日

第四章

思想和它的非凡能量

Thought and Its Extraordinary Energy

>> We have talked about the complex problem of existence, about the forming of images in our relationships with each other and the images which thought projects and which we worship. We have talked about fear, pleasure and the ending of sorrow and the question of what love is, apart from all the travail that is involved in so-called love. We have talked about compassion with its intelligence and about death. We ought now to talk about religion.

我们谈到了生活这个复杂的问题，谈到了我们在关系中互相树立形象，以及思想投射出的被我们尊崇的印象。我们谈到了恐惧、快乐和悲伤的终结，爱是什么，以及所谓的爱中牵涉的所有艰辛。我们谈到了慈悲以及它带来的智慧，谈到了死亡。我们现在应该谈谈宗教。

>> Many intellectuals, throughout the world, shy away from the subject of religion. They see what religions are in the present world, with their beliefs, dogmas, rituals and the hierarchical set-up of their established existence; and they rather scoff at and run away from anything to do with religion. And as they age and come near to that threshold called death, they often revert to their old conditioning: they become Catholics or pursue some guru in India or Japan. The more you examine, the more you are aware of the whole content of all the religious structures, the more sceptical you become about the whole business and like the intellectuals, you have nothing to do with them. And those who are not sceptical, treat religions romantically, emotionally, or as a form of entertainment.

全世界的很多知识分子都会避开宗教的话题。他们看到宗教在当前世界的面貌，及它们的信仰、教义、仪式和既存的等级机构，他们宁愿漠视和逃避任何与宗教有关的东西。而当他们变老，接近死亡的门槛时，他们常常会回归到曾经的状态，成为天主教徒，或者追随印度或日本的某个古鲁。你越是审视，越是了解宗教结构的全部内容，你就会越怀疑整件事情，就像无神论者一样，变得和它们毫不相干。而那些不怀疑的人，会以浪漫的、有情感的方式对待宗教，或者将它作为一种消遣。

>> If one puts aside the intellectual, the romantic and sentimental attitudes towards religions, one can then begin to ask, not with any naivety, but with seriousness: what is religion?—not looking for the mere meaning of that word, but deeply. Man, from ancient times, has always thought that there must be something beyond ordinary daily life, the ordinary misery, confusion and conflict of daily life. In his search he has invented all kinds of philosophies, created all

kinds of images—from those of the ancient Egyptians and the ancient Hindus to modern times—always getting caught apparently in some kind of delusion. He deludes himself and out of those delusions he creates all kinds of activities. If one could brush all that aside, not hypnotizing oneself, being free from illusion, then one can begin to examine, enquire very profoundly if there is something beyond all the contagion of thought, all the corruption of time, if there is something beyond one's usual existence in space and time and if there is any path to it, or no path, and how the mind can reach it, or come to it. If one asks that of oneself then how shall one set about it? Is any kind of preparation necessary—discipline, sacrifice, control, a certain period of preparation and then advance?

如果你撇开对宗教的知识性的、浪漫的、感性的态度，你可能会开始问，没有任何天真的想法，而是严肃地问：什么是宗教？你不是带着天真只想找到那个词的意思，而是要深入探究。人类从远古时代就一直认为，一定存在某种东西，它能够超越普通日常生活中的不幸、困惑和冲突。在寻找的过程中，人类发明了各种哲学，制造了各种形象，从古埃及和古印度直到现代的那些，而人类似乎总是陷入某种错觉。人类欺骗自己，基于那些错觉制造了所有活动。如果你能够将所有那些抛开，不对自己催眠，脱离那些幻觉，那么你就能开始非常深入地检查、探究：是否存在某种不被一切思想干扰和一切时间腐化的东西，是否存在某种超越时空中的惯常存在的东西，是否存在到达它的道路，如果说没有这样的道路，我们的心灵将如何触碰到它？如果你那样问自己，那么你该如何开始呢？准备工作是必要的吗——纪律、奉献、控制、一段时期的准备和进阶？

>> First of all, it is important to understand that one should be free of all illusions. So, what creates illusions? Is it not the desire to reach something, to experience something out of the ordinary—extrasensory perception, visions, spiritual experiences? One must be very clear as to the

nature of desire and understand the movement of desire, which is thought with its image and also have no motive in one's enquiry. It may seem very difficult to have no intention, to have no sense of direction so that the brain is free to enquire. There must be order in one's house, in one's existence, in one's relationships, in one's activity. Without order, which is freedom, there can be no virtue. Virtue, righteousness, is not something that is intellectually cultivated. Where there is order, there is virtue; that order is something that is living, not a routine, a habit.

首先要了解的是，你应该从所有幻觉中解脱出来。那么，是什么造成了幻觉？难道不是想获得什么，想体验某种非同寻常的东西的欲望吗？如超感官知觉、幻境、精神体验等等。你必须非常清楚欲望的本质，理解欲望的变化，它是一种带有意象、没有任何动机、随意的一种思想。没有目的、没有方向感似乎很难，但那样脑子才会有探寻的自由。在你的房子里，你的生活中，你的关系中，你的活动中，都要有秩序。秩序就是自由，没有秩序就不可能有美德。美德、公义不是智性的培养。有秩序的地方就有美德，那个秩序是活生生的，它不是例行公事，不是习惯。

>> Secondly, is there something to be learnt? Is there something to be learnt from another? One can learn from another, history, biology, mathematics, physics; the whole complex knowledge of the technological world one can learn from another, from books. Is there something to be learned from psychology about our lives, about that which is eternal?—if there is something eternal. Or is it that there is nothing to learn from another because all the human experience, all the psychological knowledge that humanity has gathered together for millions of years, is within oneself ? If that is so, if one's consciousness is that of the whole of mankind, then it seems rather absurd, rather naive, to try to learn from somebody else about oneself. It requires complete clarity of observation to learn about ourselves. That is simple. So there is no psychological authority and no spiritual

authority, because the whole history of mankind, which is the story of humanity, is in oneself. Therefore there is nothing to experience. There is nothing to be learnt from somebody who says: "I know" or, "I will show you the path to truth"—from the priests throughout the world, the interpreters between the highest and the lowest. Obviously, to learn about, to understand, oneself, all authority must be set aside. Authority is part of oneself; one is the priest, the disciple, the teacher, one is the experience and one is the ultimate—if one knows how to understand.

其次，需要学习什么东西吗？需要从另一个人那里学习什么东西吗？你可以从另一个人那里学习历史、生物、数学、物理，全部技术世界的复杂知识都可以从另一个人那里、从书本那里学习。而关于我们的生活，关于永恒——如果存在某种永恒的话，需要从心理学中学习什么东西吗？还是说没有任何东西需要从另一个人那里学习，因为所有人类经验、人类数百万年积累的所有的知识都存在于你自身之中？如果是这样，如果你的意识就是整个人类的意识，那么试图从别人那里学习关于自己的知识似乎就相当荒谬和幼稚了。了解我们自己需要完全清晰的观察，那很简单。因此没有心理上的权威和精神上的权威。因为人类的整个历史，即人类的故事，就在你自身之中。因而没有任何东西要去体验，没有任何东西需要从某个说“我知道”或“我将向你展示通向真理的道路”的人那里学习，包括全世界的牧师，这些居于高明者和低劣者之间的诠释者。很显然，要学习和了解自己，所有的权威都必须抛开。权威是你自己的一部分，你就是牧师、弟子、老师，你就是经验和终点，如果你知道如何去了解的话。

>> There is nothing to be learnt from anybody, including the speaker; especially one must not be influenced by the speaker. One has to be free to enquire very, very deeply, not superficially. One may have done all the superficial enquiry during the last five or fifty years, and have come to the point when one has established order, more or less, in one's life, and as one goes along one may establish greater

order, so that one can ask: what is the religious mind which can understand what meditation is?

不要从任何人那里学习什么东西，包括讲话者，尤其不要被讲话者影响。你必须自由地探究，非常深入，而不是浅尝辄止。也许你在过去的五年或五十年里进行过一切肤浅的研究，而来到了这个重要的时刻，你在自己的生活中或多或少地建立了秩序，或许在你前进的时候建立了更大的秩序，以至于你能够问：那个能够了解冥想的宗教思想是什么？

>> Within the last fifteen years, that word meditation has become very popular in the West. Before that, only very few, who had been to Asia, enquired into the Eastern forms of meditation. The Asiatics have said that only through meditation you can come to, or understand, that which is the timeless, which has no measure. But during recent years, those who have nothing to do but call themselves gurus, have come over to the West bringing that word. It

has become a word that has made meditation seem like a drug. There are also the various systems of meditation—the Tibetan, the Hindu, the Japanese Zen, and so on. These systems have been invented by thought and thought being limited the systems must inevitably be limited. And also they become mechanical, for if you repeat, repeat, your mind naturally goes dull, rather stupid and utterly gullible. It is common sense all this, but there is such eagerness to experience something spiritual, either through drugs, through alcohol, or by following a system of meditation which it is hoped will give some kind of exciting experience; This is happening; this is not exaggeration, this is not attacking anybody personally but a statement of the nonsense that is going on.

在过去的十五年里，“冥想”这个词在西方变得非常流行。在那之前，只有极少数去过亚洲的人探究过东方的冥想。亚洲人说只有通过冥想你才能实现或了解那无始无终、不可估量的东西。但是近些年来，那些无所事事却自称古鲁的人带着那个词到了西方。它让冥想看起来就像麻醉品一样。存在各种体系的冥想，中国的、印度的、日本的等等。这些体系都是思想发明的，而思想是局限的，这些体系必然也是局限的。它们会变得机械，因为如果你一再重复，你的心自然会变得迟钝、愚蠢和轻信。这一切都是常识，但是存在这种想要体验某种精神性的东西的渴望，无论是通过药物、酒精还是跟随一个冥想体系，希望它能带来某种刺激的体验。这些事情正在发生，不是夸张，不是个人化的攻击，而是对正在发生的愚蠢行为的陈述。

>> So, if one is sufficiently aware of all this, one will have put it aside, for it is utterly meaningless; and it is essential to maintain a mind that is capable, rational, sane, free from all the illusions and any form of self-hypnosis.

因此，如果充分意识到这一切，你就会把它扔掉，因为它完全没有意义。对于保持一颗有能力的、理性的、明智的心，摆脱一切幻象和任何形式的自我催眠，这是非常必要的。

>> Then what is a human being? The human being has lived on thought; all the architecture, all the music, the things that are inside the churches, the temples and mosques, they are all invented by thought. All our relationships are based on thought, though we say, "I love you", it is still based on the image which thought has created about another. Thought, to the human being, is astonishingly important; and thought itself is limited; its action is to bring about fragmentation—the fragmentation between people—my religion, my country, my belief as opposed to yours, all that is the movement of thought, space and time.

那么人是什么？人类靠思想生活，所有建筑、音乐，所有教堂、庙宇和清真寺里面的东西都是思想发明的。我们所有的关系都建立在思想上，尽管我们说“我爱你”，但是它仍然建立在思想制造的关于他人的形象之上。对人类来说，思想具有惊人的重要性，而思想本身是局限的，它的活动会带来分裂——人和人之间的分裂——我有我的宗教、我的国家、我的信仰，你有你的，所有那些都是思想、空间和时间的活动。

>> Meditation is the capacity of the brain which is no longer functioning partially—the brain which has freed itself from its conditioning and is therefore functioning as a whole. The meditation of such a brain is different from the mere contemplation of one conditioned as a Christian or a Hindu, whose contemplation is from a background, from a conditioned mind. Contemplation does not free one from conditioning. Meditation demands a great deal of enquiry and becomes extraordinarily serious in order not to function partially. By partially, it is meant to function in a particular specialization or particular occupation that makes the brain narrow in accepting beliefs, traditions, dogmas and rituals, all of which are invented by thought. Meditation is different from contemplation in the sense that meditation demands

that the brain acts wholly and is no longer conditioned to act partially. That is the requirement for meditation, otherwise it has no meaning.

冥想能让大脑不再局部运作，可以让大脑从自身的制约中解脱出来，因而能够整体地运作。这样的脑子产生的冥想不同于一个基督徒或印度教徒的观想，他们的观想有一个背景，来自一颗被制约的心。观想不会把你从约束中解脱出来。冥想需要大量的探究和极端认真，才能让大脑不陷入局部的运作。局部的意思是指以一种特别的专业化分工或特别的职业方式去运作，从而使脑子狭隘地接受信仰、传统、教条和仪式，那一切都是思想发明的。冥想不同于观想，就这个意义来说，冥想需要脑子完整地运作，而不是被约束着进行局部的运作。那是冥想所需要的，否则它就没有意义。

>> So the question is: is it possible to live in this world, which demands certain forms of specialization, a skilful mechanic, mathematician, or housewife, yet to be free from

specialization? Suppose I am a theoretical physicist and have spent most of my life in mathematical formulation, thinking about it, questioning it, cultivating considerable knowledge about it, so that my brain has become specialized, narrowed down and then I begin to enquire into meditation. Then in my enquiry into meditation I can only partially understand the significance and the depth of it because I am anchored in something else, in the theoretical physics of my profession; anchored there I begin to enquire theoretically whether there is meditation, whether there is the timeless; so my enquiry becomes partial again. But I have to live in this world; I am a professor at a university; I have a wife and children, I have that responsibility and perhaps I am also ill; yet I want to enquire very profoundly into the nature of truth, which is part of meditation. So the question is: is it possible to be specialized as a theoretical physicist and yet leave it at a certain level so that my brain (the brain which is the common brain of all humanity) can say: yes, it has that specialized function but that function is not going to interfere?

所以问题是，有没有可能活在这个需要某种形式的专业化分工，需要熟练的技工、数学家或家庭主妇的世界上，还能从专业化分工中解脱出来？假设我是一个理论物理学家，消耗了生命中大部分的时间来进行数学表达，思考它，质疑它，培养很多关于它的知识，以至于我的脑子变得非常专业、狭隘，于是我开始探索冥想。在我探索冥想的过程中，我只能部分地了解冥想的含义和深度，因为我还紧抓着别的事情，抓着我的理论物理专业；同时，我开始从理论上探索是否存在冥想，是否存在永恒；于是我的探索又成了局部的。但是我必须活在这个世界上，我是一个大学教授，我有妻子和孩子，我有那些责任，或许我还有病，然而我非常热切地想探索真理的本质，这就是冥想的一部分。所以问题是：有没有可能作为一个理论物理学家，同时又停留在某个层面上，从而让我的脑子（整个人类共有的脑子）能够意识到“是的，它有专门的作用，但是这个作用不会产生障碍”？

>> If I am a carpenter, I know the quality of the wood, the grain, the beauty of the wood and the tools with which to work it. And I see that that is natural and I also see that the brain that has cultivated the speciality cannot possibly understand the wholeness of meditation. If as a carpenter I understand this, the truth of it, that I, as a carpenter have a place, but also that that specialization has no place in the wholeness of comprehension, in the wholeness of understanding meditation, then that specialization becomes a small affair.

如果我是一个木匠，我了解木材的品质、纹理和木头的美，以及用来加工它的工具。我看到这些是自然的事情，而我也能看到培养这些专长的脑子不可能了解冥想的整体。如果我作为木匠了解这一点，了解它的真相，即我作为木匠有一个位置，但是这种专业化分工在对整体性的领悟中、在对冥想的整体性了解中是没有位置的，那么这个专业化分工就变成了一件不重要的事情。

>> So then we begin to ask: what is meditation? First of all, meditation demands attention, which is to give your whole capacity, energy, in observation. Attention is different from concentration. Concentration is an effort made by thought to focus its capacity, its energy, on a particular subject. When you are in school, you are trained to concentrate, that is to bring all your energy to a particular point. In concentration you are not allowing any other kind of thoughts to interfere; concentration implies the controlling of thought, not allowing it to wander away but keeping it focused on a particular subject. It is the operation of thought which focuses attention, focuses energy, on that subject. In that operation of thought there is compulsion, control. So in concentration there is the controller and the controlled. Thought is wandering off ; thought says it should not wander off, and I bring it back as the controller who says, "I must concentrate on this." So there is a controller and the controlled. Who is the controller? The controller is part of thought and the controller is the past. The controller

says, “I have learnt a great deal and it is important for me, the controller, to control thought.” That is: thought has divided itself as the controller and the controlled; it is a trick that thought is playing upon itself. Now, in attention there is no controller, nor is there the controlled, there is only attention. So a careful examination is required into the nature of concentration with its controller and the controlled. All our life there is this controller—“I must do this, I must not do that, I must control my desires, control my anger, control my impetus.”

所以我们开始问：冥想是什么？首先，冥想需要注意，就是用你全部的才能、精力去观察。注意不是专注，专注是一种把思想的能力和能量聚焦到一个特定主题上的努力。你在上学的时候受到训练去专注，那是要将你的所有精力带到一个特定的点。专注的时候不允许其他思想干预，专注意味着对思想的控制，不允许它到处漫游，而是将它投到一个特定的主题上面。它是把注意力和精力聚集在主题上的思想的运作。在那个思想的运作中存在着强迫和控制。因此专注当中有控制者和被控制者。思想在徘徊，思想说它不该徘徊，作为控制者我把它带回来，我说："我必须专注在这个上面。"因此存在控制者和被控制者。控制者是谁？控制者是思想的一部分，控制者是过去。控制者说："我学了很多东西，控制者、控制思想对我来说很重要。"也就是说，思想将自己分割为控制者和被控制者，这是一个思想玩弄自己的把戏。而注意当中没有控制者，也没有被控制者，只有注意。因此需要对专注、对控制者和被控制者进行仔细检查。我们的全部生活中都存在这样的控制——"我必须做这个，我必须控制我的欲望、控制我的愤怒、控制我的冲动。"

>> We must be very clear in understanding what concentration is and what attention is. In attention there is no controller. So, is there in daily existence, a way of living in which every form of psychological control ceases to exist?—because control means effort, it means division between the controller and the controlled; I am angry, I must control my

anger; I smoke, I must not smoke and I must resist smoking. We are saying there is something totally different and this may be misunderstood and may be rejected altogether because it is very common to say that all life is control—if you do not control you will become permissive, nonsensical, without meaning, therefore you must control. Religions, philosophies, teachers, your family, your mother, they all encourage you to control. We have never asked: who is the controller? The controller is put together in the past, the past which is knowledge, which is thought. Thought has separated itself as the controller and the controlled. Concentration is the operation of that. Understanding that, we are asking a much more fundamental question, which is: can one live in this world, with a family and responsibilities, without a shadow of control?

在了解专注是什么、注意是什么的过程中，我们的头脑必须非常清晰。注意之中没有控制者。因此，在我们的日常生活中，存不存在一种生活方式，其中每一种形式的心理控制都不存在了？因为控制意味着努力，意味着在控制者和被控制者之间进行划分。我发怒了，我必须控制我的愤怒；我吸烟，而我绝不能吸烟，我必须抵制吸烟。我们说存在某种完全不同的东西，这也许会被误解进而被彻底抛弃，因此说生活全部都需要控制是很平常的——如果你不控制你就会变得放任、荒谬、没有意义，因而你必须控制。宗教、哲学、老师、你的家庭、你的母亲，他们都鼓励你控制。我们从来不问控制者是谁，控制者是在过去拼凑起来的，过去就是知识，就是思想。思想把自己分割成控制者和被控制者。专注就是那样运作的。了解它之后，我们要问一个更为基本的问题，那就是：你能否带着家庭和责任活在这个世界上，而没有控制的影子？

>> See the beauty of that question. Our brain has been trained for thousands of years to inhibit, to control, and now it is never operating with the wholeness of itself. See for yourself what it is doing; watch your own brain in operation, rationally, critically examining it in a way in which there is

no deception or hypnosis. Most of the meditations that have been put forward from the Asiatic world involve control; control thought so that you have a mind that is at peace, that is quiet, that is not eternally chattering. Silence, quietness and the absolute stillness of the mind, the brain, are necessary in order to perceive and to achieve these forms of meditation, however subtle, have control as their basis. Alternatively you hand yourself over to a guru, or to an ideal and you can forget yourself because you have given yourself over to something and therefore you are at peace, but again it is the movement of thought, desire and the excitement of attaining something you have been offered.

要看到这个问题的美。我们的脑子数千年来都被训练去抑制、控制，从来没有整体地运作。亲自看看它在做什么，密切注意自己脑子的运作，以一种理性的、批判性的方式检查它，没有欺骗或催眠。大多数来自亚洲的冥想都包含控制，控制思想以使你获得一颗和平、安静、不再躁动的心。为了理解和实现这一点，内心和脑子的沉默、安静和彻底止息是必要的。这些形式的冥想，无论多么微妙，都是以控制为基础的。把自己交给一个古鲁或一种理想，可以使你忘掉自己，因为你把自己交出去了，所以你平静了，但那还是思想和欲望的活动，是因为获得某种东西而产生的兴奋。

>> Attention is not the opposite of concentration. The opposite has its root in its own opposite. If love is the opposite of hate, then love is born out of hate. Attention is not the opposite of concentration, it is totally divorced from it. Does attention need effort? That is one of our principal activities; I must make an effort; I am lazy, I do not want to get up this morning, but I must get up, make an effort. I do not want to do something but I must. See how extraordinary it is that we cannot catch the significance of this immediately. It has to be explained, explained, explained. We seem to be incapable of direct perception of the difference between concentration and attention; unable to have an insight into attention and be attentive.

注意不是专注的反面。反面根植于自己的正面之中。如果爱是恨的反面，那么爱就来自恨。注意不是专注的反面，它是完全不同的。注意需要努力吗？努力是我们的主要活动之一，我必须努力，我很懒，今天早上我不想起床，但是我必须起床，做出努力。我不想做某事，但是我必须做。看看这多么离奇，我们无法马上理解它的意义。因此它必须被一再解释。我们似乎不能直接觉察专注和注意之间的差别，不能了解注意和留心。

>> When does attention take place? Obviously not through effort. When one makes an effort to be attentive, it is an indication that one is inattentive and is trying to make that inattention become attention. See instantly that the observer is the observed and therefore one makes no effort, it is so. Effort exists when there is division. Does it not indicate that one's brain has become dull because one has been trained, trained, so it has lost its pristine quickness, its capacity to see directly without all the explanations

and words, words, words. But unfortunately one has to go into this because one's mind, one's brain, cannot, for example grasp instantly, that truth has no path; it is unable to see the immensity of that statement, the beauty of it and put aside all paths so that one's brain becomes extraordinarily active. One of the difficulties is that one has become mechanical. If one's brain is not extraordinarily alive and active, it will gradually wither away. Now one's brain has to think, it has to be active, if only partially, but when the computer can take over all the work and most of the thought, operating with a rapidity which the brain cannot, then the brain is going to wither. This is happening, it is not an exaggerated statement of the speaker, it is happening now and we are unaware of it.

什么时候注意会发生？显然不是通过努力。当你努力去注意的时候，表明你是不注意的，并试图将那个不注意变成注意。要立即看到观察者同时也是被观察者，因而不再努力，就是这样。努力存在于分裂中。它表明你的脑子迟钝了，因为你不断被训练，让它失去了原始的锐利，失去了不需要言语解释而直接看事物的能力。但不幸的是，你必须深入了解这一点，你的心、你的脑子不能马上理解“并没有一条通往真理的道路”，你因无法看到这句话的广大和它的美，从而放弃寻路，但这样一来，你的脑子却变得出奇地活跃。困难之一是你已经变得机械了。如果你的脑子不是出奇地敏锐和活跃，它就会渐渐枯萎。现在你的脑子必须思考，必须保持活跃，计算机能够以脑子不可能达到的极快速度运作，接替了所有工作和大多数思考的时候，脑子就会枯萎。这正在发生，不是讲话者夸张地陈述，它现在正在发生，而我们却没有注意。

>> In concentration there is always a centre from which one is acting. When one concentrates, one is concentrating for some benefit, for some deep-rooted motive; one is observing from a centre. Whereas in attention there is no centre at all. When one looks at something immense—like the mountains with their extraordinary majesty, the line

against the blue sky and the beauty of the valley—the beauty of it for a moment drives out the centre; one is for a second stunned by the greatness of it. Beauty is that perception when the centre is not. A child, given a toy, is so absorbed by it that he is no longer mischievous, he is completely with the toy. But he breaks the toy and he is back to himself. Most of us are absorbed by our various toys; when the toys go, we are back to ourselves. In the understanding of ourselves without the toy, without any direction, without any motive, is the freedom from specialization which makes the whole of the brain active. The whole of the brain when it is active is total attention.

专注时总是有一个关注点，你基于这个关注点行动。你专注是因为某种利益，因为某种深刻的动机，是从一个关注点出发来观察的。然而注意之中根本没有关注点。当你看某种巨大的东西，例如山脉在蓝天的映衬下曲折蜿蜒以及它不可思议的庄严，还有山谷的美，它们瞬间就驱散了那个关注点，片刻间你被它们的伟大震惊了。美是没有关注点的觉察。一个得到玩具的孩子，完全被它所吸引，以至于不再淘气，和玩具玩得不亦乐乎。但是如果玩具坏了，他就又做回了自己。我们大多数人都会被各种玩具吸引，当玩具不在的时候，我们又重新做回自己。在没有玩具、没有任何方向、没有任何动机的情况下，对自身的了解让我们从专业化中解脱出来，让整个脑子变得活跃。当整个脑子生机勃勃的时候，我们就处于完整的注意中了。

>> One is always looking or feeling with part of the senses. One hears some music, but one never really listens. One is never aware of anything with all one's senses. When one looks at a mountain, because of its majesty, one's senses are fully in operation, therefore one forgets oneself. When one looks at the movement of the sea or the sky with the slip of a moon, when one is aware totally, with all one's senses, that is complete attention in which there is no centre. Which means that attention is the total silence of the brain, there is no longer chattering, it is completely still—an

absolute silence of the mind and the brain. There are various forms of silence—the silence between two noises, the silence between two notes, the silence between thoughts, the silence when you go into a forest—where there is the great danger of a dangerous animal, everything becomes totally silent. This silence is not put together by thought, nor does it arise through fear. When one is really frightened one's nerves and brain become still—but meditation is not that quality of silence, it is entirely different. Its silence is the operation of the whole of the brain with all the senses active. It is freedom which brings about the total silence of the mind. It is only such a mind, such a mind-brain, that is absolutely quiet—not quietness brought about by effort, by determination, by desire, by motive. This quietness is the freedom of order, which is virtue, which is righteousness in behaviour. In that silence alone is there that which is nameless and timeless. That is meditation.

26th July, 1981

你总是使用部分感官去看或感觉。你听到一段音乐，但是你从来没有真正去听。你从未用全部的感官去觉察什么。当你看一座山的时候，因为它非常雄伟，你的感官充分运作，于是你忘记了自己。当你看到大海翻涌，看到月亮爬过天空，当你用自己全部的感官完整地觉察，就产生了完整的注意，其中没有关注点。那意味着注意是脑子彻底静默，不再喋喋不休，完全地静止——一种心和脑的绝对静默。存在各种形式的静默：两个声音之间的静默，两个音符之间的静默，思想之间的静默，你走进森林面对野兽的巨大威胁时的完全静默。这种静默不是思想组成的，也不是由恐惧产生的。当你真正害怕的时候，你的神经和脑子就停止了——但是冥想不是那样的静默，它是完全不同的。冥想是整个脑子以及所有生机勃勃的感官的运作。是自由导致了脑子的整体静默。只有这样的心、这样的心脑整体，才拥有绝对的安静，不是由努力、决心、欲望、动机带来的安静。这种安静是有秩序的自由，这就是美德，就是义行。

在那份静默之中本身就有那不可名状、无始无终之物。那就是冥想。

一九八一年七月二十六日

>> Most unfortunately there are only two talks and so it is necessary to condense what we have to say about the whole of existence. We are not doing any kind of propaganda; we are not persuading you to think in one particular direction, nor convince you about anything—we must be quite sure of that. We are not bringing something exotic from the East like the nonsense that goes on in the name of the gurus and those people who write strange things after visiting India—we do not belong to that crowd at all. And we would like to point out that during these two talks we are thinking together, not merely listening to some ideas and either agreeing or disagreeing with them; we are not creating arguments, opinions, judgements, but together—I mean together, you and the speaker—we are going to

observe what the world has become, not only in the West, but also in the East. We cannot therefore rely on any politicians, on any leaders, or on any books that have been written about religion. We cannot possibly rely on any of these people, nor on the scientists, the biologists, or the psychologists. They have not been able to solve our human problems. I am quite sure you agree to all that.

非常遗憾，只有两次谈话了，我们不得不简化我们要说的，有关生活的全部内容。我们不是在做任何宣传，不是要劝你朝一种特定的方向思考，也不是想让你相信什么，对此我们完全确定。我们并不是从东方带来了一种奇特的东西，就像某些人正在以古鲁的名义胡说八道，以及某些人在造访印度之后写一些奇闻逸事，我们完全不属于那类人。我们想要指出的是，在这两次谈话中，我们是在一同思考，而不只是听到一些想法，对它们表示同意或不同意。我们不是在制造争论、意见、评判，而是你和讲话者在一起，我们将观察世界成了什么样子，不只是在西方，也包括东方。我们不能完全依靠政治家、领导或谈论宗教的书籍，也不能完全依靠科学家、生物学家或心理学家。他们没能解决人类的全部问题。我相当确定你会同意所有这些。

>> Having said all this it is important that we, you and the speaker, think together. We mean by thinking together not merely accepting any kind of opinion or evaluation but observing together, not only externally what is happening in the world, but also what is happening to all of us inwardly, psychologically. Externally, outwardly, there is great uncertainty, confusion, wars, or the threat of war. There are wars going on now in some parts of the world; human beings are killing each other. That is not happening in the West, here, but there is the threat of nuclear war, and the preparation for war.

说到这里我想再强调一遍，重要的是我们，你和讲话者，一同思考。我们说一同思考的意思是，不只是接受任何意见或判断，而是一同观察，不只是观察外部世界正在发生的事情，还要观察我们所有人心理上正在发生的事情。外在有着巨大的不确定、混乱、战争或战争的威胁。现在世界上的一些地方正在发生战争，一些人类在互相残杀。它没有发生在西方，但是这里存在着核战争的威胁，以及为战争所做的准备。

>> And inwardly, in our own minds and in our own hearts, we ourselves are also very confused. There is no security, not only, perhaps, for ourselves but for the future generation. Religions have divided human beings as the Christians, the Hindus, the Muslims, and the Buddhists. So considering all this, observing objectively, calmly without any prejudice, it is naturally important that together we think about it all. Think together, not having opinions opposing other sets of opinions, not having one conclusion against another conclusion, one ideal against another ideal, but rather

thinking together and seeing what we human beings can do. The crisis is not in the economic world, nor in the political world; the crisis is in consciousness. I think very few of us realize this. The crisis is in our mind and in our heart; that is, the crisis is in our consciousness. Our consciousness is our whole existence. With our beliefs, with our conclusions, with our nationalism, with all the fears that we have, it is our pleasures, the apparently insoluble problems and the thing that we call love, compassion; it includes the problem of death—wondering if there is anything hereafter, anything beyond time, beyond thought and if there is something eternal: that is the content of our consciousness.

在我们的头脑中和心里，我们自己也非常困惑。不仅我们自己没有安全感，或许未来的一代也没有。宗教把人们划分成基督教徒、印度教徒、穆斯林和佛教徒。考虑到这一切，客观、平静、没有任何成见的观察，以及我们所有人一同来思考自然就成了非常重要的事情。一同思考，不是一些意见反对另一些意见，一个结论反对另一个结论，一个理想反对另一个理想，而是一同思考，看看我们人类能做什么。危机不在经济领域或政治领域，它在意识里。我想我们当中几乎没有人能认识到这点。危机就在我们的头脑中和心里。换句话说，危机在我们的意识里面。我们的意识是我们的全部。意识伴随着信仰、结论、民族主义、我们所有的恐惧，它还是我们的愉悦之源、我们似乎无法解决的问题和那个被我们称为爱和慈悲的东西。它还包含死亡的问题——不知道有没有来世，有没有超越时间、思想的东西，某种永恒的东西。那些就是我们的意识。

>> That is the content of the consciousness of every human being, in whatever part of the world he lives. The content of our consciousness is the common ground of all humanity. I think this must be made very clear right from the beginning. A human being living in any part of the world suffers, not only physically but also inwardly. He is uncertain, fearful, confused, anxious without any sense of deep security. So

our consciousness is common to all mankind. Please do listen to this. You may be hearing this for the first time, so please do not discard it. Let us investigate it together, let us think about it together, not when you get home but now: your consciousness, what you think, what you feel, your reactions, your anxiety, your loneliness, your sorrow, your pain, your search for something that is not merely physical but goes beyond all thought, is the same as that of a person living in India or Russia or America. They all go through the same problems as you do, the same problems of relationship with each other, man, woman. So we are all standing on the same ground of consciousness. Our consciousness is common to all of us and therefore we are not individuals. Please do consider this. We have been trained, educated, religiously as well as scholastically, to think that we are individuals, separate souls, striving for ourselves, but that is an illusion because our consciousness is common to all mankind. So we are mankind. We are not separate individuals fighting for ourselves. This is logical, this is rational, sane. We are not separate entities with separate psychological content, struggling for ourselves, but we are, each one of us, actually the rest of humankind.

那是每一个人的意识，无论他生活在世界的什么地方。我们人类的意识是共通的。我想这点必须从一开始就弄清楚。在这个世界上任何地方生活的人都会痛苦，不管生理上的还是心理上的。他不确定、恐惧、困惑、焦虑、没有任何安全感。因此我们的意识是整个人类共有的。请注意听这些话。你也许第一次听到这些，请不要抛弃它。让我们共同研究它，共同思考它，就在现在，不是在你回家以后：你的意识、想法、感受，你的反应、忧虑、孤独、悲伤、痛苦，你对某种不只是物质的而是超越一切思想的东西的探索，你拥有的这些和一个生活在印度或俄罗斯或美国的人拥有的是相同的。他们和你一样都经历了同样的问题，同样的人际关系问题，无论是男人还是女人。因此我们都站在相同的意识基础上。我们的意识是共通的，因而我们不是个人。请认真考虑这件事。我们在宗教中和学校里受到训练和教育，由此认为我们是个体，是分开的灵魂，只为自己奋斗，但那是一个假象，因为我们的意识是全人类共通的。我们不是分离的为自己拼争的个体。这是合乎逻辑的，这是理性的、明智的。我们不是具有独立心理的为自己拼争的独立个体。但实际上我们每一个人都和其余的人一样。

>> Perhaps you will accept the logic of this intellectually, but if you feel it profoundly then your whole activity undergoes a radical change. That is the first issue we have to think about together: that our consciousness, the way we think, the way we live, some perhaps more comfortably, more affluently, with greater facility to travel than others, is inwardly, psychologically, exactly similar to that of those who live thousands and thousands of miles away.

或许你理智上可以接受这个逻辑，但是如果你深切地感受到它，那么你的整个活力就会发生根本性的改变。这是我们必须一同考虑的首要问题：我们心理上的意识、思维方式、生活方式——或许有些人比别人更舒适，更富裕，进行更多旅行——和那些生活在千里之外的人是相同的。

>> All is relationship, our very existence is to be related. Observe what we have done with our relationships with each other, whether intimate or not. In all relationship there is tremendous conflict, struggle—why? Why have human beings, who have lived for over a million years, not solved this problem of relationship? So let us this morning think together about it. Let us observe together actually what the relationship between a man and a woman is. All society is based on relationship. There is no society if there is no relationship, society then becomes an abstraction.

一切都是关系，我们就是通过关系联系起来的。观察我们在彼此关系中的所作所为，无论是否亲密。所有的关系中都存在着巨大的冲突、斗争，为什么？为什么存在了一百万年以上的人类没有解决关系的问题？

所以今天上午让我们共同来思考它。让我们一同实际地观察一个男人和一个女人之间的关系到底是什么。一切社会都建立在关系上。如果没有人际交往就没有社会，社会就会变成一个抽象的概念。

>> One observes that there is conflict between man and woman. The man has his own ideals, his own pursuits, his own ambitions, he is always seeking success, to be somebody in the world. And the woman is also struggling, also wanting to be somebody, wanting to fulfil, to become. Each is pursuing his or her own direction. So it is like two railway lines running parallel, never meeting, but otherwise—if you observe closely—never actually meeting psychologically, inwardly. Why? That is the question. When we ask why, we are always asking for the cause; we think in terms of causation, hoping that if we could understand the cause then perhaps we would change the effect.

男人和女人之间存在冲突。男人有他自己的理想、追求、野心，他总在追求成功，想成为这个世界上数一数二的人物。女人也在奋斗，也想成为什么人，想要有成就。每个人都在追求他或她自己的方向。他们就像两条平行的铁轨，永远不会相交，除了在床上。如果你密切观察的话，会发现他们从来没有在心理上相遇过。为什么？这是个问题。当问为什么的时候，我们总是会去找原因，我们根据原因来思考，希望通过了解原因来改变结果。

>> So we are asking a very simple but very complex question: why is it that we human beings have not been able to solve this problem of relationship though we have lived on this earth for millions of years? Is it because each one has his own particular image put together by thought, and that our relationship is based on two images, the image that the man creates about her and the image the woman creates about him? So in this relationship we are as two images living together. That is a fact. If you observe yourself very closely,

if one may point out, you have created an image about her and she has created a picture, a verbal structure, about you, the man. So relationship is between these two images. These images have been put together by thought. And thought is not love. All the memories of this relationship with each other, the pictures, the conclusions about each other, are, if one observes closely without any prejudice, the product of thought; they are the result of various remembrances, experiences, irritations and loneliness, and so our relationship with each other is not love but the image that thought has put together. So if we are to understand the actuality of relationships, we have to understand the whole movement of thought, because we live by thought; all our actions are based on thought, all the great buildings, the cathedrals, churches, temples and mosques of the world are the result of thought. And everything inside these religious buildings—the figures, the symbols, the images—are all the invention of thought. There is no refuting that. Thought has created not only the most marvellous buildings and the contents of those buildings, but it has also created the instruments of war, the bomb in all its various forms. Thought has also produced the surgeon and his marvellous instruments, so delicate in surgery. And thought has also produced the carpenter, his study of wood and the tools he uses. The contents of a church, the skill of a surgeon, the expertise of the engineer who builds a beautiful bridge, are all the result of thought—there is no refuting that. So one has to examine what thought is and why human beings live on thought and why thought has brought about such chaos in the world—war and lack of relationship with each other—and examine the great capacity of thought with its extraordinary energy. We must also see how thought has, through millions of years, brought such sorrow for mankind. Please observe this together, let us examine it together. Do not just oppose what the speaker is saying, but examine what he is saying together so that we understand what is actually happening to all of us human beings.

我们要问一个非常简单而又非常复杂的问题：为什么我们人类没有能够解决关系的问题，尽管我们已经在这个地球上生活了上百万年？是因为每一个人都有他自己经由思想组织起来的特别的形象，我们的关系都建立在两个形象之上，建立在男人建立的关于女人的形象和女人建立的关于男人的形象之上吗？在这种关系中，我们是作为两个形象生活在一起的，这是一个事实。如果可以指出的话，那就是你非常密切地观察自己，然后你制造了一个关于她的形象，她制造了一个关于你的图像和词汇。所以关系存在于这两个形象之间。思想把这些形象整合在一起。思想并不是爱。关于这段关系的所有记忆、形象以及对彼此的判断才是爱。如果可以不加偏见地仔细观察的话，就会明白思想的产物是各种回忆、经历、愤怒、孤独共同作用的结果。关系亦是如此，它并非来自爱，而是来自思想拼绘在一起的形象。因此如果我们要了解关系的现实，那么就必须了解思想的整个活动，因为我们靠思想活着。我们的所有行动都建立在思想上面，世界上所有宏伟的建筑，教堂、庙宇和清真寺都是思想的结果。这些宗教建筑里面的一切，图画、符号、形象，都是思想发明的。无可否认，思想不仅制造了最精妙的建筑和它里面的东西，还制造了战争工具，如各种样式的炸弹。思想催生了外科医生和他不可思议的仪器，使他在手术中表现得如此微妙娴熟。思想也催生了木匠、木匠对木材的研究和他使用的工具。教堂里的东西、外科医生的技术、建造美丽桥梁的工程师的专业知识，都是思想的产物，这无可否认。因此你必须检视思想是什么，为什么人类活在思想中，为什么思想给这个世界带来了那么多混乱——战争，彼此关系的缺失，用非凡的能量检视思想的巨大能力。我们也必须看到，在数百万年的时间中思想如何给人类带来悲伤。请一同观察这些，让我们一同检视它。不要只是反对讲话者说的东西，而是一同检视他的话，从而理解我们所有人身上真实发生的是什么。

>> Thought is the response of the memory of things past; it also projects itself as hope into the future. Memory is knowledge; knowledge is memory of experience. That is, there is experience, from experience there is knowledge

as memory, and from memory you act. From that action you learn, which is further knowledge. So we live in this cycle—experience, memory, knowledge, thought and thence action—always living within the field of knowledge.

思想是对过去记忆的反应，它也把自己作为希望投射到将来。记忆是知识，知识是对经验的记忆。也就是说，先有经验，对经验的记忆形成了知识，并且你基于记忆来行动。你从那个行动中学习，那又进一步形成了知识。因此我们活在这个循环中：经验、记忆、知识、思想，然后行动，我们总是活在知识的领域中。

>> What we are talking about is very serious. It is not something for the weekend, for a casual listening, it is concerned with a radical change of human consciousness. So we have to think about all this, look together, and ask why we human beings, who have lived on this earth for so many millions of years, are still as we are. We may have advanced technologically, have better communication, better

transportation, hygiene and so on, but inwardly we are the same, more or less—unhappy, uncertain, lonely, carrying the burden of sorrow endlessly. And any serious man confronted with this challenge must respond; he cannot take it casually, turn his back on it. That is why these meetings are very, very serious because that is why we have to apply our minds and our hearts to finding out if it is possible to bring about a radical mutation in our consciousness and therefore in our action and behaviour.

我们谈论的这些是非常严肃的。它不是一次随意的周末讲座，它关系到人类意识的彻底转变。所以我们必须思考这一切，共同去看，并问：为什么在地球上生活了数百万年的人类，现在依然是这个样子？我们也许在技术上取得了进步，有了更好的通信设备，更好的交通设施、卫生条件等等，但是心理上我们或多或少还是一样，不幸、不确定、孤独，无休止地背负着悲伤的重担。任何面对这个严肃挑战的人都必须做出回答，他不能对它漫不经心，转过身去。那就是为什么这些聚会是非常非常严肃的，因为我们必须运用我们的头脑和我们的心去探明，这些有没有可能给我们的意识，进而给我们的行动和行为带来根本性的转变。

>> Thought is born of experience and knowledge, and there is nothing sacred whatsoever about thought. Thinking is materialistic, it is a process of matter. And we have relied on thinking to solve all our problems in politics and religions and in our relationships. Our brains, our minds, are conditioned, educated to solve problems. Thinking has created problems and then our brains, our minds, are trained to solve them with more thinking. All problems are created, psychologically and inwardly, by thought. Follow what is happening. Thought creates the problem, psychologically; the mind is trained to solve problems with further thinking, so thought in creating the problem then tries to solve it. So it is caught in a continuous process, a routine. Problems are becoming more and more complex, more and more insoluble, so we must find out if it is at all possible to

approach life in a different way, not through thought because thought does not solve our problems; on the contrary thought has brought about greater complexity. We must find out—if it is possible or not—whether there is a different dimension, a different approach, to life altogether. And that is why it is important to understand the nature of our thinking. Our thinking is based on remembrance of things past—which is thinking about what happened a week ago, thinking about it modified in the present, and projected into the future. This is actually the movement of our life. So knowledge has become all-important for us but knowledge is never complete. Therefore knowledge always lives within the shadow of ignorance. That is a fact. It is not the speaker's invention or conclusion, but it is so.

思想来自经验和知识，没有什么神圣可言。思想是物质性的，它是一个物质过程。我们依靠思想来解决政治的、宗教的和关系中的问题。我们的脑子，我们的心被制约、教育成解决问题的模式。思想制造了问题，然后我们的脑子、我们的心又被训练用更多的思想去解决它们。所有心理上的问题都是思想制造的。请跟上我说的东西。思想在心理上制造了问题，理智受到训练通过进一步的思想去解决，是那个制造问题的思想试图去解决它。因此它陷入了一个连续的过程，一个程序当中。问题变得越来越复杂，越来越无法解决。所以我们必须搞清楚，到底有没有可能以一种不同的方式处理生活的问题，不是通过思想，因为思想不能解决我们的问题。相反，它带来了更大的复杂性。我们必须弄清楚，对于完整的生活而言，是不是存在一种完全不同的维度和方式。那就是为什么了解我们思想的本质是非常重要的事情。我们的思想建立在对过去的回忆上面，它总在考虑一周以前发生的事情，被现在修正过的事情，以及投射到将来的事情。这就是我们真实的生活。因此知识变得对我们非常重要，但是知识从来都是不完整的，知识总是伴随着无知。这是一个事实，不是讲话者的发明或结论，它就是如此。

>> Love is not remembrance. Love is not knowledge. Love is not desire or pleasure. Remembrance, knowledge, desire and pleasure are based on thought. Our relationship with each other, however near, if looked at closely, is based on remembrance, which is thought. So that relationship—though you may say you love your wife or your husband or your girlfriend—is actually based on remembrance, which is thought. And in that there is no love. Do you actually see that fact? Or do you say, "What a terrible thing to say. I do love my wife"—but is that so? Can there be love when there is jealousy, possessiveness, attachment, when each one is pursuing his own particular direction of ambition, greed and envy, like two parallel lines never meeting? Is that love?

爱不是记忆，不是知识，不是欲望或快乐。记忆、知识、欲望和快乐都建立在思想上面。如果你密切观察的话，你会发现，我们之间的关系无论多么亲近，都是建立在记忆，也就是思想上面的。因此那种关系——尽管你或许会说你爱你的妻子或你的丈夫或你的女朋友——实际上都建立在记忆，也就是思想上，其中没有爱。你能真正看到这个事实吗？或者你说："你说得太可怕了，我确实爱我妻子。"是那样的吗？当存在妒忌、占有、依附的时候，当两个人就像两条永不相交的平行线，都在追求自己特殊的野心、贪婪和嫉妒的时候，爱可能存在吗？那是爱吗？

>> I hope we are thinking together, observing together, as two friends walking along a road and seeing what is around us, not only what is very close and immediately perceived, but what is in the distance. We are taking the journey together, perhaps affectionately, hand in hand—two friends amicably examining the complex problem of life, neither of their leader or guru, because when one sees actually that our consciousness is the consciousness of the rest of mankind, then one realizes that one is both the guru and the disciple, the teacher as well as the pupil, because all that is in one's consciousness. That is a tremendous realization. So as one begins to understand oneself deeply one becomes a light to

oneself and not dependent on anybody, on any book or on any authority—including that of the speaker—so that one is capable of understanding this whole problem of living and of being a light to oneself.

我希望我们是在一起思考，一起观察，就像两个朋友一起散步，一起看到周围的东西，不仅能马上发觉近处的东西，还能发现远处的东西。我们是在一同旅行，或许可以满怀深情、密切合作——两个朋友友善地检视复杂的生活问题，谁都不是领导者或古鲁，因为当你真正去看的时候，我们的意识就是其他人类的意识，那么你就会了解，你既是古鲁也是弟子，既是老师也是学生，因为所有那些都在你的意识里面。那是一种巨大的领悟。因此当你开始深入了解自己的时候，你就变成了照亮自己的光，无须依赖任何人、书籍或权威——包括讲话者的权威，这样你才能够理解整个生活的问题，才能成为自己的光。

>> Love has no problems and to understand the nature of love and compassion with its own intelligence, we must understand together what desire is. Desire has extraordinary vitality, extraordinary persuasion, drive, achievement; the whole process of becoming, success, is based on desire—desire which makes us compare ourselves with each other, imitate, conform. It is very important in understanding the nature of ourselves to understand what desire is, not to suppress it, not to run away from it, not to transcend it, but to understand it, to see the whole momentum of it. We can do that together, which does not mean that you are learning from the speaker. The speaker has nothing to teach you. Please realize this. The speaker is merely acting as a mirror in which you can see yourself. Then when you see yourself clearly, you can discard the mirror, it has no more importance, you can break it up.

爱没有问题，要了解爱、慈悲以及其智慧的本质，我们就必须一同了解欲望是什么。欲望有着非同一般的重要性，非同一般的说服力、驱动力和成就，一切的成长和成功都建立在欲望的基础上，它让我们相互比较、模仿、遵从。要了解自己的本质，了解欲望是非常重要的，不是压制它、逃离它、超越它，而是了解它，看到它的全部。我们可以一同做这件事，这不意味着你在向讲话者学习。讲话者没有什么东西要教你，请了解这一点。他只是像镜子一样，你可以在其中看到自己。然后当你清晰地看到自己的时候，你就能够抛弃镜子，它不再重要了，你可以打破它。

>> To understand desire requires attention, seriousness. It is a very complex problem to understand why human beings have lived on this extraordinary energy of desire as on the energy of thought. What is the relationship between thought and desire? What is the relationship between desire and will? We live a great deal by will. So what is the movement, the source, the origin, of desire? If one observes oneself one sees the origin of desire; it begins with sensory responses; then thought creates the image and at that

moment desire begins. One sees something in the window, a robe, a shirt, a car, whatever it is—one sees it, sensation, then one touches it, and then thought says,"If I put on that shirt or dress how nice it will look"—that creates the image and then begins desire. So the relationship between desire and thought is very close. If there were no thought there would only be sensation. Desire is the quintessence of will. Thought dominates sensation and creates the urge, the desire, the will, to possess. When in relationship thought operates—which is remembrance, which is the image created about each other by thought—there can be no love. Desire, sexual or other forms of desire, prevent love—because desire is part of thought.

了解欲望需要注意和认真。了解人类为什么要靠欲望拥有的不寻常的能量活着，就像人类依靠思想的能量活着一样，这是一个非常复杂的问题。思想和欲望的关系是什么？欲望和意志的关系是什么？我们在很大程度上靠意志活着。欲望的活动、根源、起点是什么？

如果观察自己，你就会看到欲望的起点。它开始于感官反应，然后思想制造了印象，此刻欲望就开始了。你在橱窗里看到了什么，一件长袍，一件衬衫，一辆汽车，无论是什么——你看到它、感受它、碰触到它，然后思想说“要是我穿上那件衬衫或衣服，看上去会很漂亮”——它都制造了印象，然后欲望就开始了。因此欲望和思想的关系非常紧密。如果没有思想，那么就只有感知。欲望是意志的本质。思想支配感官并催生了驱动力、欲望、意志。当思想在关系中运作——也就是记忆，是思想制造的对彼此的印象——就不可能有爱。欲望、性或其他形式的欲望，都阻碍了爱，因为欲望是思想的一部分。

>> We should consider in our examination the nature of fear because we are all caught in this terrible thing called fear. We do not seem to be able to resolve it. We live with it, become accustomed to it, or escape from it through amusement, through worship, through various forms of entertainment, religious and otherwise. Fear is common to

all of us, whether we live in this tidy, clean country, or in where it is untidy, dirty and overpopulated. It is the same problem, fear, which man has lived with for thousands and thousands of years and which he has not been able to resolve. Is it possible—one is asking this question most seriously—is it at all possible to be totally, completely, free of fear, not only the physical forms of fear but the much more subtle forms of inward fear—conscious fears and the deep undiscovered fears which we have never even known were there? Examination of these fears does not mean analysis. It is the fashion to turn to the analyst if you have any problem. But the analyst is like you and me, only he has a certain technique. Analysis implies there is an analyser. Is the analyser different from that which he analyses? Or is the analyser the analysed? The analyser is the analysed. That is an obvious fact. If I am analysing myself, who is the analyser in me who says, "I must analyse"? It is still the analyser separating himself from the analysed and then examining that which is to be analysed. So the analyser is that which he is analysing. They are the same. To separate them is a trick played by thought. But when we observe, there is no analysis; there is merely the observing of things as they are—the observing of that which actually is, not analysing that which is, because in the process of analysing we can deceive ourselves. If you like to play that game you can, and go on endlessly until you die, analysing, and never bringing about a radical transformation within yourself. Whereas to look at the present as it is—not as a Dutchman, an Englishman, or a Frenchman or as this or that—to see what is actually happening, is pure observation of things as they are.

在检视的时候，我们应当考虑恐惧的本质，因为我们都陷在这个东西当中。我们似乎不能解决它，我们忍受它，变得习惯它，通过娱乐、崇敬、各种形式的消遣、宗教或其他方式来逃避它。无论我们生活在整洁的国家，还是生活在不整洁、肮脏、人口过剩的国家，恐惧都是我们所有人共通的。它是共同的问题，人类和它共处了成千上万年，却从来没能解决它。一个人极为认真地提出这个问题：到底有没有可能完全、彻底地从恐惧中解脱出来？不仅是生理上的恐惧，还有更加微妙的心理上的恐惧——有意识的恐惧，深藏不露的，我们甚至从来都不知道它的存在的恐惧。检视这些恐惧并不意味着分析。在有问题的时候求助于精神分析师成了一种时尚。但是分析师和你我一样，只不过他拥有某种技巧。分析意味着有一个分析者。分析者和他分析的问题有什么不同吗？还是说分析者就是被分析的对象？分析者就是被分析者。那是显而易见的事实。如果我分析自己，那个在我脑海里面说“我必须分析”的分析者是谁？仍旧是那个把自己和被分析者分离，然后检查被分析者的分析者。因此分析者就是他在分析的东西，它们是相同的。分离它们是思想玩的一个把戏。但是我们观察的时候不存在分析者。只有对事物如实观察，对实际发生的东西进行观察，而不是分析，因为在分析的过程中我们能够欺骗自己。如果你喜欢，你也可以没完没了地玩那个游戏一直到死，不停地分析，虽然它永远不能带来内在的根本转变。如实地看现在，看实际发生的东西，不是作为荷兰人、英国人、法国人或这样那样的人去看，而是对事物进行如实的纯然的观察。

>> To observe what fear is, is not to examine the cause of fear, which implies analysis and going further and further back into the origin of fear. It is to learn the art of observing and not translating or interpreting what you observe, but just observing, as you would observe a lovely flower. The moment you take it to pieces, the flower is not. That is what analysis does. But observe the beauty of a flower, or the evening light in a cloud, or a tree by itself in a forest, just observe. So similarly, we can observe fear and what is the root of fear—not the various aspects of fear.

观察什么是恐惧，不是去检查恐惧的原因，那意味着分析，那和探究恐惧的根源渐行渐远。是去学习观察的艺术，不去解释你观察到的东西，只是观察，就像你观察一朵可爱的花。在你扯碎它的时候花就不美了，那就是分析所做的事情。观察一朵花、傍晚的霞光、或森林里中独自矗立的树木，观察它的美，只是观察。类似地，我们也可以观察恐惧和它的根源，而不是不同种类的恐惧。

>> We are asking if it is at all possible to be free of fear, absolutely. Psychologically, inwardly, what is the root of fear? What does fear mean? Does not fear arise from something that has given you pain in the past which might happen again in the future? Not what might happen now because now there is no fear. You can see for yourself that fear is a time process. Something that happened last week, an incident which brought psychological or physical pain, and from that there is fear that it might happen again tomorrow. Fear is a movement in time; a movement from the past through the present, modifying the future. So the origin of fear is thought. And thought is time, it is the accumulation of knowledge through experience, the response of memory as thought, then action. So thought and time are one; thought and time are the root of fear. That is fairly obvious. It is so.

我们要问，到底有没有可能从恐惧中彻底解脱。在心理上，恐惧的根源是什么？恐惧意味着什么？恐惧难道不是从那些过去给你造成痛苦、将来也许还会发生的事情中产生的吗？不是现在可能发生的事情，因为现在不存在恐惧。你可以意识到，恐惧的产生需要时间。上周发生了一个事件，导致了心理上和生理上的痛苦，由此产生了“它也许明天还会发生”的恐惧。恐惧是时间中的活动，它来自过去，经过现在，改变着将来。因此恐惧的根源就是思想。而思想就是时间，它是通过经验、思想对记忆的反应和行动而累积知识。因此思想和时间是一个东西，思想和时间就是恐惧的根源，相当明显，就是这样。

>> Now it is not a question of stopping thought or time. Of course it would be impossible to stop them because the entity who says, “I must stop thought” is part of thought. So the idea of stopping thought is absurd. It implies a controller who is trying to control thought and such a controller is created by thought. Please just observe this; observation is an action in itself, it is not that one must do something about fear. I wonder if you understand this?

现在并不是停止思想或时间的问题。停止它们当然是不可能的，因为那个说“我必须停止思想”的主体就是思想的一部分。因此停止思想的想法是荒谬的，它隐含着一个试图控制思想的控制者，这样的控制者是思想制造的。请只是观察这点，观察本质上就是行动，而不是说你必须对恐惧做些什么。我想知道你们是否理解这些。

>> Suppose I am afraid about something or other, darkness, my wife running away, loneliness, or this or that. I am frightened, deeply. You come along and explain to me the whole movement of fear, the origin of fear, which is time. I had pain; I went through some accident or incident that caused pain, that is recorded in the brain, and the memory of that past incident produces the thought that it might happen again, and therefore there is fear. So you have explained this to me. And I have listened very carefully to your explanation, I see the logic of it, the sanity of it, I do not reject it; I listen. And that means that listening becomes

an art. I do not reject what you are saying, nor accept, but I observe. And I observe that what you tell me about time and thought, is actual. I do not say, "I must stop time and thought", but having had it explained to me, I just observe how fear arises, that it is a movement of thought, time. I just observe this movement and do not move away from it, I do not escape from it but live with it, look at it, put my energy into looking. Then I see that fear begins to dissolve because I have done nothing about it, I have just observed, I have given my whole attention to it. That very attention is like bringing light on fear. Attention means giving all your energy in that observation.

假设我害怕某件事情，害怕黑暗，怕我妻子跑掉，怕孤独，怕这个或那个，我深深地恐慌。你过来向我解释恐惧的整个活动，恐惧的根源，它就是时间。我曾痛苦过，我经历的一些事情导致了痛苦，脑子记录了它们，对过去事件的记忆导致了“它也许还会发生”的思想，因而产生了恐惧。因此你对我解释这些。我非常仔细地听你的解释，我看到其中的逻辑，它的合理性，我不拒绝它，我在听。那意味着这个听变成了艺术。我不拒绝你说的东西，也不接受，只是观察。我观察到你告诉我的关于时间和思想的事情是实际的。我不会说“我必须停止时间和思想”，而是在听到这个解释之后，观察恐惧是如何发生的，了解它是思想、时间的活动。我只是观察这个活动，不离开它，不逃避它，与它共处，看着它，投入我的能量去看。然后我看到恐惧开始溶解，因为我对它什么都没做，我只是观察，对它付出我的全部注意。正是那份注意带来了照亮恐惧的光。注意意味着付出你全部的能量去观察。

>> Why is it that man pursues pleasure? Please ask yourself why. Is pleasure the opposite of pain? We have all had pain of different kinds, both physical and psychological. Psychologically, most of us from childhood have been wounded, hurt; that is pain. The consequence of that pain has been to withdraw, to isolate oneself so as not

to be further hurt. From childhood, through school, by comparing ourselves with somebody else who is more clever, we have hurt ourselves, and others have hurt us through various forms of scolding, saying something brutal, terrorizing us. And there is this deep hurt with all its consequences, which are isolation, resistance, more and more withdrawal. And the opposite of that we think is pleasure. Pain and the opposite of it is pleasure. Is goodness the opposite of that which is not good? If goodness is the opposite, then that goodness contains its own opposite. Therefore it is not good. Goodness is something totally separate from that which is not goodness. So is pleasure something opposite to pain? Is it a contrast? We are always pursuing the contrast, the opposite. So one is asking, is pleasure entirely separate, like goodness, from that which is not pleasurable? Or is pleasure tainted by pain? When you look closely at pleasure it is always remembrance, is it not? You never say when you are happy, "How happy I am", it is always after; it is the remembrance of that which gave you pleasure, like a beautiful sunset, the glory of an evening, full of that extraordinary light, it gave great delight. Then that is remembered and pleasure is born. So pleasure is part of thought too—it is so obvious.

为什么人类要追求快乐？请问问你自己为什么。快乐是痛苦的反面吗？我们都有过各种不同的痛苦，包括生理上和心理上的。心理上，我们大多数人从童年时代就受到过伤害，那就是痛苦。痛苦的结果就是退缩，孤立自己，为了不再受到进一步的伤害。从童年时代起，在学校，通过和别人比较谁更聪明，我们伤害了自己；其他人则通过各种形式的斥责、无情的语言、威胁来伤害我们。始终存在这些深刻的伤害，及其所有的后果，即孤立、抗拒、越来越退缩。我们认为它们的反面是快乐，痛苦的反面是快乐。善是恶的反面吗？如果善是恶的反面，那么善就包含自己的反面，那么它就是不善。善是某种和不善完全不同的东西。快乐是和痛苦相反的东西吗？当你密切地看快乐，它总是记忆，不是吗？你快乐的时候从来不说“我好快乐”，说总是在你感到快乐之后发生。它是对带给你快乐的事物的回忆，就像美丽的晚霞、壮丽的黄昏、绚烂的光，它带给我们狂喜。然后我们回忆它，快乐就产生了。因此快乐也是思想的一部分，这如此明显。

>> The understanding of relationship, fear, pleasure and sorrow, is to bring order in our house. Without order you cannot possibly meditate. Because there is no possibility of right meditation if you have not put your house, your psychological house, in order. If the psychological house is in disorder, if what you are is in disorder, what is the point of meditating? It is just an escape. It leads to all kinds of illusions. You may sit cross-legged or stand on your head for the rest of your life, but that is not meditation. Meditation must begin with bringing about complete order in your house—order in your relationships, order in your desires, pleasures and so on.

对关系、恐惧、快乐和悲伤的了解，会为我们的“房子”带来秩序。没有秩序你就不可能冥想。因为如果你没有让你的“房子”，你心理上的“房子”变得有秩序，那么就不可能有正确的冥想。如果心理上的“房子”处于混乱之中，如果你自身处于混乱之中，那么冥想的核心要点是什么呢？那只是一种逃避，它会导致所有幻觉。你或许在你的余生都盘腿坐着，但是那不是冥想。冥想必须从在你的“房子”里实现完全的秩序开始——你关系中的秩序，你的欲望、快乐等之中的秩序。

>> One of the causes of disorder in our life is sorrow. This is a common factor, a common condition, in all human beings. Everyone goes through this tragedy of sorrow, whether in the Asiatic world or in the Western world. Again this is a common thing we all share. There is not only so-called personal sorrow, but there is the sorrow of mankind, the sorrow which wars have brought about—five thousand years historical records and every year there has been a war, killing, violence, terror, brutality, the maiming of people,

people who have no hands, no eyes—the horrors and the brutality of wars which have brought incalculable misery to mankind. It is not only one's own sorrow but the sorrow of mankind; the sorrow of seeing a man who has nothing whatsoever, just a piece of cloth, and for the rest of his life he is going to be that way. And when you see that person there is sorrow. There is also sorrow when people are caught in illusion, like going from one guru to another, escaping from themselves. It is a sorrow to observe this, the clever people going off to the East, writing books about it, finding some guru—so many fall for that nonsense. There is the sorrow that comes when you see what the politicians are doing in the world—thinking in terms of tribalism. There is personal sorrow and the vast cloud of the sorrow of mankind. Sorrow is not something romantic, sentimental, illogical; it is there. We have lived with this sorrow from time measureless, and apparently we have not resolved this problem. When we suffer we seek consolation, which is an escape from the fact of sorrow. When there is that grief, you try every form of amusement and escape, but it is always there. Apparently humanity has not resolved it. And we are asking the question: is it possible to be free of it completely? Not avoiding it, not seeking consolation, not escaping into some fanciful theory, but is it possible to live with it? Understand those words "to live with it" : they mean not to let sorrow become a habit. Most people live with sorrow, they live with their own separate religious conclusions, they live with their own fanciful ideas and ideals, which all again bring conflict. So live with something, live with sorrow, not accepting it, not becoming habituated to it—but look at it, observe it without any escape, without any question of trying to go beyond it, just "hold it in your hand" and look. Sorrow is also part of the tremendous sense of loneliness: you may have many friends, you may be married, you may have all kinds of things, but inwardly there is this feeling of complete loneliness. And that is part of sorrow. Observe that loneliness without any direction, without trying to go beyond it, without trying to find a substitute for it; live with it, not worship it, not become psychotic about it, but give all your attention to that loneliness, to that grief, to that sorrow.

我们生活混乱的原因之一是悲伤。这是所有人的共同基因、共同状况。每个人都会经历悲伤，无论在亚洲还是在西方世界。这也是我们所有人共同承担的东西。不仅有所谓的个体悲伤，还有人类整体的悲伤，例如战争带来的悲伤——五千年的历史记录显示，每一年都会有战争、杀戮，以及暴力、恐怖、野蛮事件发生，致使很多人变成没有手、没有眼睛的残疾人——战争的恐怖和残暴给人类带来了无尽的痛苦。那不仅是你自己的悲伤，也是人类的悲伤。看到一个人衣衫褴褛，余生将会一直如此，你就会感到悲伤。还有当人们陷入幻觉时的悲伤，他们从信仰一个古鲁换成信仰另一个古鲁，不停地逃避自己。观察到这些也是一种悲伤：聪明的人们去东方，写关于东方的书，找到一些古鲁，那么多人听信了这些无意义的东西。基于部落意识思考，当你看到政治家在世界上的所作所为时，会产生悲伤。有个人的悲伤，也有人类的广大的悲伤。悲伤不是某种浪漫的、情绪化的、合乎逻辑的东西，它就在那里。我们自始至终都和悲伤一起生活，很显然我们没有解决这个问题。在受苦的时候我们寻求安慰，那是对悲伤现实的逃避。在悲伤的时候，你试图进行各种娱乐和逃避，但是它总在那里。显然人类没有解决它。我们要问这个问题：有没有可能彻底从悲伤中解脱出来？不逃避，不寻求安慰，不躲进某种虚幻的理论中，有没有可能和它共处？了解“和它共处”这句话：它意味着不要让悲伤变成习惯。大多数人都在忍受悲伤，他们忍受自己分裂的宗教论断，忍受自己虚幻的想法和理想，那些都会带来冲突。因此试着和事物共处，和悲伤共处，不是接受它、习惯它，只是看着它，没有任何逃避，没有任何超越它的打算，只是“将它握在手里”然后去看。悲伤也是可怕的孤独感的一部分。你可能有很多朋友，你也许结婚了，也许拥有很多东西，但是内心存在着这种完全孤独的感受。那是悲伤的一部分。观察那种孤独，没有任何方向，不试图去超越它，或找个东西替代它，和它共处，不去崇拜它，不对它神魂颠倒，而是对那种孤独、那种悲伤付出你全部的注意。

>> It is a great thing to understand suffering because where there is freedom from sorrow there is compassion. Compassion is freedom from sorrow. Where there is compassion there is love. With that compassion goes intelligence—not the intelligence of thought with its cunning, with its adjustments, with its capacity to put up with anything. Compassion means the ending of sorrow and only then is there intelligence.

19th September, 1981

了解痛苦是一件伟大的事情，因为从悲伤中解脱会带来慈悲。慈悲就是从悲伤中解脱而来。有慈悲就会有爱。智慧会伴随着慈悲到来，而非来自狡猾的、多变的、能适应一切的思想。慈悲意味着悲伤的终结，那时智慧才会到来。

一九八一年九月十九日

第五章

狡猾的思想

Network of Thought

>> We are like two friends sitting in the park on a lovely day talking about life, talking about our problems, investigating the very nature of our existence, and asking ourselves seriously why life has become such a great problem, why? Though intellectually we are very sophisticated, yet our daily life is such a grind, without any meaning, except survival—which again is rather doubtful. Why has life, everyday existence, become such a torture? We may go to church, follow some leader, political or religious, but the daily life is always a turmoil, though there are certain periods which are occasionally joyful, happy, there is always a cloud of darkness about our life. And these two friends, as we are, you and the speaker, are talking over

together in a friendly manner, perhaps with affection, with care, with concern, whether it is at all possible to live our daily life without a single problem. Although we are highly educated, have certain careers and specializations yet we have these unresolved struggles, the pain and suffering, and sometimes joy and a feeling of not being totally selfish.

我们就像两个朋友，在一个美好的日子里，坐在公园里探讨生活，探讨我们的问题，研究我们存在的本质，并且认真地问我们自己，为什么生活成了这样一个巨大的问题，为什么？尽管理智上我们非常成熟，然而我们的日常生活是如此让人疲倦，没有任何意义，除了生存之外——那也相当难以预料。为什么我们每天的生活成了这样一种折磨？我们可以去教堂，去追随某个政治上或宗教上的领袖，但是日常生活总是一片混乱，尽管某个时期会有偶尔的喜悦、愉快。两个朋友，你和讲话者，在以一种友好的方式相互讨论，或许可以带着情感、关心、在意，究竟我们能否没有任何问题地过每天的生活。尽管我们受过高等教育，拥有某种职业和专门技能，但仍拥有这些未解决的挣扎、伤害和痛苦，以及间或的喜悦和不是完全自私的感觉。

>> So let us go into this question of why we human beings live as we do, going to the office from nine until five or six for fifty years, and always the brain, the mind, constantly occupied. There is never a quietness, but always this occupation with something or other. And that is our life. That is our daily, monotonous, rather lonely, insufficient life. And we try to escape from it through religion, through various forms of entertainment. At the end of the day we are still where we have been for thousands and thousands of years. We seem to have changed very little, psychologically, inwardly. Our problems increase, and always there is the fear of old age, disease, some accident that will put us out. So this is our existence, from childhood until we die, either

voluntarily or involuntarily die. We do not seem to have been able to solve that problem, the problem of dying. Especially as one grows older one remembers all the things that have been the times of pleasure, the times of pain, and of sorrow, and of tears. Yet always there is this unknown thing called death of which most of us are frightened. And as two friends sitting in the park on a bench, not in this hall with all this light, which is rather ugly, but sitting in the dappling light, the sun coming through the leaves, the ducks on the canal and the beauty of the earth, let us talk this over together. Let us talk it over together as two friends who have had a long serious life with all its trouble, the troubles of sex, loneliness, despair, depression, anxiety, uncertainty, a sense of meaninglessness—and at the end of it always death.

让我们深入地探讨这个问题：为什么人类要像现在这样生活，五十年如一日地从早上九点到下午五点或六点到办公室上班，脑子、心一直不断地被占据。从来没有一刻安静，总是被什么东西占据着。那就是我们每天所过的单调枯燥、相当孤独和贫乏的生活。我们试图通过宗教、通过各种形式的娱乐逃避它。当一天结束的时候我们仍然待在原地，那个我们待了许多年的地方。在心理上我们似乎改变了一点。我们的问题增多了，并一直有着对衰老、疾病、打扰我们的事情的恐惧。这就是我们的生活，从童年开始直到死亡，要么坦然地死，要么抗拒地死。我们似乎没能解决死亡的问题。特别是当一个人变老的时候，他回想起所有经历的时光，所有的快乐、痛苦、悲伤、泪水的时候。总是存在这个未知的叫作死亡的东西，我们大多数人都害怕它。就像两个朋友坐在公园的长椅上，不是在这个光芒四射的大厅里，而是坐在斑驳的光影里，阳光透过树叶照射下来，鸭子在水渠里畅游，大地很美，我们一起谈论这件事。让我们像两个朋友一样一起去谈论它，我们经历了漫长的严肃生活，其中充满着各种各样的麻烦，包括性、孤独、绝望、沮丧、忧虑、不确定、无意义感，最终是死亡。

>> In talking about it, we approach it intellectually—that is, we rationalize it, say it is inevitable, not to fear it or, if you are highly intellectual, telling yourself that death is the end of all things, of our existence, our experiences, our memories, be they tender, delightful, plentiful; the end also of pain and suffering. What does it all mean, this life which is really, if we examine it very closely, rather meaningless? We can, intellectually, verbally, construct a meaning to life, but the way we actually live has very little meaning. Living and dying is all we know. Everything apart from that is theory, speculation; meaningless pursuit of a belief in which we find some kind of security and hope. We have ideals projected by thought and we struggle to achieve them. This is our life, even when we are very young, full of vitality and fun, with the feeling that we can do almost anything; but with youth, middle and old age supervening, there is always this question of death.

通过谈论这件事，我们可以在智力上接近它，就是说，将它合理化，谈论它是不可避免的，不要恐惧它；或者如果你智力超群，告诉自己死亡是一切事物的终结；我们的存在、我们的经验、我们的记忆，它们是温柔的、令人愉快的、丰富的，也是伤害和痛苦的终结。这意味着什么？这所谓的真实生活，如果我们认真地审视它，就会发现它其实毫无意义。我们能够在理智上、口头上建立一种生活的意义，但是我们实际的生活却没有意义。生活和死亡是我们知道的一切。除此之外的一切都是理论、推测，对信仰的无意义追求，我们只是在其中寻找某种安全和希望。我们拥有思想投射出来的理想，我们通过奋斗去实现它。这就是我们的生活，甚至在我们非常年轻的时候就是如此，但那时我们充满了活力和兴趣，感觉我们几乎能做任何事情。而死亡的问题总是在青年、中年和老年时存在。

>> You are not merely, if one may point out, listening to a series of words, to some ideas, but rather together, I mean

together, investigating this whole problem of living and dying. And either you do it with your heart, with your whole mind, or else partially, superficially—and so with very little meaning.

如果可以指出的话，你不只是在听一系列的言辞、想法，而是在与我一同调查生活和死亡的问题。要么用你的心、你的整个头脑去做，要么部分地、表面化地去做，但这样没有什么意义。

>> First of all we should observe that our brains never act fully, completely; we use only a very small part of our brain. That part is the activity of thought. Being in itself a part, thought is incomplete. The brain functions within a very narrow area, depending on our senses, which again are limited, partial; the whole of the senses are never free, awakened. I do not know if you have experimented with watching something with all your senses, watching the sea, the birds and the moonlight at night on a green lawn, to see

if you have watched partially or with all your senses fully awakened. The two states are entirely different. When you watch something partially you are establishing more the separative, egotistically centred attitude to living. But when you watch that moonlight on the water making a silvery path with all your senses, that is with your mind, with your heart, with your nerves, giving all your attention to that observation, then you will see for yourself that there is no centre from which you are observing.

首先我们应该观察，我们的脑子从来都不能充分、完整地运作，我们只是用了其中很小的一部分。那部分是思想的活动。作为整体中的一个部分，思想是不完全的。脑子依赖我们的感官，在一个狭隘的范围内运作，它也是受限的、局部的。整个感官从来不是自由的、觉醒的。我不知道你是否尝试过，用你的所有感官去注视一个东西，注视大海、鸟儿、夜晚草地上的月光，去看看你是局部地观察，还是用你全部的完全觉醒的感官观察。这两种状态是完全不同的。当你局部地观察一件事情，你建立了一种分裂的、以自我为中心的生活态度。但是当你用全部的感官，也就是用你的头脑、你的心、你的神经，付出你全部的注意，去观察在水面上铺泻银色道路的月光，那么你就会看到，并没有一个中心是你所观察的。

>> Our ego, our personality, our whole structure as an individual, is entirely put together from memory; we are memory. Please, this is subject to investigation, do not accept it. Observe it, listen. The speaker is saying that the "you", the ego, the "me", is altogether memory. There is no spot or space in which there is clarity—you can believe, hope, have faith, that there is something in you which is uncontaminated, which is god, which is the spark of that which is timeless, you can believe all that, but that belief is merely illusory. All beliefs are. But the fact is that our whole existence is entirely memory, remembrances. There is no spot or space inwardly which is not memory. You can investigate this; if you are enquiring seriously

into yourself you will see that the “me”, the ego, is all memory, remembrances. And that is our life. We function, live, from memory. And for us, death is the ending of that memory.

我们的自我，我们的人格，我们作为个体的整个结构，完全是由记忆构成的，我们就是记忆。拜托，这是需要探究的主题，不要接受它。要观察它，倾听。讲话者说“你”、自我、“我”完全都是记忆。没有一个地方或空间是清晰的。你可以相信、怀抱希望、怀抱信仰，你心里有某种未被污染的东西，它是永恒的火焰，你可以相信那一切，但是那个信仰只是虚无缥缈的，所有的信仰都是如此。事实是，我们的存在完全是记忆、回忆。内心没有什么地方不是记忆。你可以探究这件事。如果你仔细探究自己，你就会看到，“我”、自我全都是记忆、回忆。那就是我们的生活。我们是基于记忆运作和生活的。对我们来说，死亡就是记忆的终结。

>> Am I speaking to myself or are we all together in this? The speaker is used to talking in the open, under trees, or in a vast tent without these glaring lights, then we can have an intimate communication with each other. As a matter of fact there is only you and I talking together, not this enormous audience in a vast hall, but you and I sitting on the banks of a river, on a bench, talking over this thing together. And one is saying to the other, we are nothing but memory, and it is to that memory that we are attached—my house, my property, my experience, my relationship, the office or the factory I go to, the skill I like being able to use during a certain period of time—I am all that. To all that, thought is attached. That is what we call living. And this attachment creates all manner of problems; when we are attached there is fear of losing; we are attached because we are lonely with a deep abiding loneliness which is suffocating, isolating, depressing. And the more we are attached to another, which is again memory, for the other is a memory, the more problems there are. I am attached to the name, to the

form; my existence is attachment to those memories which I have gathered during my life. Where there is attachment I observe that there is corruption. When I am attached to a belief, hoping that in that attachment there will be a certain security, both psychologically as well as physically, that attachment prevents further examination. I am frightened to examine when I am greatly attached to something, to a person, to an idea, to an experience. So corruption exists where there is attachment. one's whole life is a movement within the field of the known. This is obvious. Death means the ending of the known. It means the ending of the physical organism, the ending of all the memory which I am, for I am nothing but memory—memory being the known. And I am frightened to let all that go, which means death. I think that is fairly clear, at least verbally. Intellectually you can accept that logically, sanely; it is a fact.

我是在对自己讲话，还是我们大家一同在讨论？讲话者习惯在露天讨论，在树底下，或者在一个巨大的帐篷里，如果没有这些耀眼的光，那么我们相互之间就能够有一种密切的交流。事实上只有你和我在一起谈话，没有其他在场的听众，只有你和我，坐在河边长椅上，一同谈论这件事。一个人对另一个人说，除了记忆，我们什么也不是，我们依附那些记忆——我的房子、我的财产、我的经验、我的关系、我去的办公室或工厂，一段时期内我能够运用的技能——我就是那一切。思想对那一切都在依附。那就是我们所谓的生活。这种依附制造了各式各样的问题。当我们依附的时候，我们就会有失去的恐惧。因为我们带着深刻而持久的孤独，它令人窒息、隔绝、沮丧，所以我们依附。我们越是依附另一个人——那是又一个记忆，因为别人就是一个记忆——就会有越多问题。我们依附名字、形式，我的存在就是对那些生活中累积的记忆的依附。我观察到，有依附的地方就有腐化。当我依附一个信仰，希望那个依附带来心理上和生理上的安全感，那个依附就妨碍了进一步的检视。当我严重依附一个人、一个想法、一个经验的时候，我会害怕去检视。因此腐化存在于有依附的地方。你整个生活就是一个已知领域里的活动。这很明显。死亡意味着已知的终结，它意味着有机体的终结，所有记忆也就是“我”的终结，因为除了作为已知的记忆，我什么也不是。我害怕失去那一切，那意味着死亡。我认为这相当清楚，至少口头上是。理智上你可以接受它是合逻辑的、合理的。这是一个事实。

>> Now the question is: is it possible, while one is living, with all the energy, capacity and turmoil, to end, for example, attachment? Because that is what is going to happen when you die. You may be attached to your wife or husband, to your property. You may be attached to some belief in god which is merely a projection, or an invention, of thought, but you are attached to it because it gives a certain feeling of security however illusory it is. Death means the ending of that attachment. Now while living, can you end voluntarily, easily, without any effort, that form of attachment? Which means dying to something you have known—you follow? Can you do this? Because that is dying together with living,

not separated by fifty years or so, waiting for some disease to finish you off. It is living with all your vitality, energy, intellectual capacity and with great feeling, and at the same time for certain conclusions, certain idiosyncrasies, experiences, attachments, hurts to end, to die. That is, while living, also live with death. Then death is not something far away, death is not something that is at the end of one's life, brought about through some accident, disease or old age, but rather an ending to all the things of memory—that is death, a death not separate from living.

那么问题是：当一个人带着所有精力、才能和混乱活着的时候，有没有可能停止像依附这样的事情？因为停止依附是在你死去的时候才能发生的事情。你或许依附你的妻子或丈夫，依附你的财产，依附于对神明的信仰，但那只是一种思想的投射，或捏造，你依附它是因为它给你带来了某种安全感，无论它多么虚幻。死亡意味着依附的终结。那么在活着的时候，你能不能自发地、轻易地、毫不努力地停止那种依附？你已知的一些东西，你追随的东西死去了——你能承受吗？因为那是活着时即死去，而不是把生死分开五十几年，等待某种疾病把你清除。它是伴着你的所有活力、能量、智力、才能和巨大的感受的生活，同时还有某些结论、癖好、经验、依附、伤害的终结、死亡。也就是说，在活着的时候，也带着死亡。那么死亡就不是某种遥远的东西，不是一个人生活的终结，不是某种事故、疾病或年老带来的东西，而是记忆中所有事情的终结——这就是死亡，一种和生活不曾分离的死亡。

>> Also we should consider as two friends sitting together on the banks of a river, with the clear water flowing, seeing the movement of the waves pursuing each other down the river, why religion has played such a great part in people's lives from the most ancient of times until today? What is a religious mind, what is it like? What does the word "religion" actually mean? Because historically civilizations have disappeared, and new beliefs have taken their place, which have brought about new civilizations and new cultures—not

the technological world of the computers, the submarines, the war materials, nor the businessmen, nor the economists, but religious people throughout the world have brought about a tremendous change. So one must enquire together into what we mean by "religion". What is its significance? Is it mere superstition, illogical and meaningless? Or is there something far greater, something infinitely beautiful? To find that, is it not necessary to be free of all the things which thought has invented about religion?

作为朋友，我们一同坐在河边，伴着清澈的流水，看着顺流而下、互相追逐的波浪，我们也应该考虑，为什么从最远古的时代直到今天，宗教在人们的生活中扮演了重大的角色？什么是一颗宗教性的心，它是什么样的？“宗教”这个词实际上意味着什么？因为历史的文明消失了，新的信仰取而代之，这带来了新的文明和新的文化——不是技术世界的计算机、潜水艇、战争物资，也不是商人，不是经济学家，而是全世界的宗教人士带来的一种巨大的改变。因此你必须一同探究“宗教”意味着什么。它的意义是什么？它只是没有逻辑、没有意义的迷信吗？或者存在更为巨大的东西，无限美好的东西？当然要去发现它，难道我们没有必要摆脱思想制造的关于宗教的一切东西吗？

>> Man has always sought something beyond the physical existence. He has always searched, asked, suffered, tortured himself, to find out if there is something which is not of time, which is not of thought, which is not belief or faith. To find that out one must be absolutely free, for if you are anchored to a particular form of belief, that very belief will prevent investigation into what is eternal—if there is such a thing as eternity which is beyond all time, beyond all measure. So one must be free—if one is serious in the enquiry into what religion is—one must be free of all the things that thought has invented about that which is considered religious. That is, all the things that Hinduism, for example, has invented,

with its superstitions, with its beliefs, with its images, and its ancient literature such as the Upanishads—one must be completely free of all that. If one is attached to all that, then it is impossible, naturally, to discover that which is original. You understand the problem? But all religions, whether Christian, Muslim, Hindu, Buddhist, are the movement of thought continued through time, through literature, through symbols, through things made by the hand or by the mind—and all that is considered religious in the modern world. To the speaker that is not religious. To the speaker it is a form of illusion, comforting, satisfying, romantic, sentimental but not actual.

人类一直在寻找某种超越物质世界的东西。为了探明是否存在某种不属于时间、不属于思想、不是信念或信仰的东西，人类一直在探索、询问、忍耐、扭曲自己。要探明它，你必须绝对自由，因为如果你抓紧一种特定的信念，那个信念就会妨碍对永恒的探索，如果存在这种超越一切时间、一切度量的永恒的事物的话。

因此你必须是自由的，如果你认真探究宗教是什么，你必须免于思想发明的一切被认为是宗教的东西。比如印度教发明的一切东西，它的迷信，它的信仰，它的形象，它古老的文献，你必须彻底摆脱那一切。如果你依附所有那些，那么要发现那最初的自然就是不可能的。你了解这个问题吗？但是所有的宗教，无论是基督教徒、穆斯林、印度教徒、佛教徒，都是思想通过时间、文学、符号、手和大脑制造的东西延续的活动，那一切在现代社会被认为是宗教。对讲话者来说那不是宗教。对讲话者来说，它是一种幻觉，给人安慰，让人满意，它浪漫、多愁善感而又不切实际。

>> We live in disorder—that is, in conflict, contradiction, saying one thing, doing another, thinking one way and acting in another way; that is contradiction. Where there is contradiction, which is division, there must be disorder. And a religious mind is completely without disorder. That is the foundation of a religious life—not all the nonsense that

is going on with the gurus with their idiocies.

我们活在混乱中，也就是说，活在冲突、矛盾中，说一样做一样，想是一回事，行动是另一回事，那就是矛盾。有矛盾，也就是有分裂的地方，就一定有混乱。一颗宗教性的心完全没有混乱，那是宗教生活的基础，而不是那些古鲁们正在进行的所有胡说和愚行。

>> It is a most extraordinary thing how many gurus have come to see the speaker, some of them because they think I attack them. They want to persuade me not to attack, they say what you are saying and what you are living is the absolute truth, but it is not for us because we must help those people who are not as fully advanced as you are. You see the game they play—you understand? So one wonders why some Western people go to India, follow these gurus, get initiated—whatever that may mean—put on different robes and think they are very religious. But strip them of their

robes, stop them and enquire into them, and they are just like you and me.

最离奇的事情是，有那么多古鲁来看讲话者，其中一些人是因为他们认为我在攻击他们。他们想要说服我不要攻击，他们说你的言论和你的生活绝对是真理，但是它不适合我们，因为我们必须帮助那些没有像你那样充分进步的人。看看他们玩的把戏，你了解吗？所以我奇怪为什么一些西方人去印度，跟随古鲁，获得传授——无论那意味着什么——穿上不同的长袍，认为他们非常具有宗教性。可是脱下他们的长袍，拦住并探究他们，你会发现，他们只是和你我一样的人。

>> So the idea of going somewhere to find enlightenment, of changing your name to some Sanskrit name, seems strangely absurd and romantic, without any reality—but thousands are doing it. Probably it is a form of amusement without

much meaning. The speaker is not attacking. Please let us understand that: we are not attacking anything, we are just observing—observing the absurdity of the human mind, how easily we are caught; we are so gullible.

因此，去某个地方寻求开悟，把你的名字改成梵文的想法似乎非常荒谬和怪诞，没有任何现实意义，但是成千上万的人在那样做。或许它是一种没有什么意义的消遣。讲话者不是在攻击谁，要了解这点：我们不是在攻击任何事情，我们只是在观察，观察人类的荒谬，我们多么容易被操控，多么容易受骗。

>> A religious mind is a very factual mind; it deals with facts, with what is actually happening with the world outside and the world inside. The world outside is the expression of the world inside; there is no division between the outer and the inner. A religious life is a life of order, diligence, dealing with that which is actually within oneself, without any illusion, so that one leads an orderly, righteous life.

When that is established, unshakeably, then we can begin to enquire into what meditation is.

宗教性的心是非常实际的，它依据事实，依据外在和内在世界实际发生的事情来运作。外在世界是内在世界的表现，外在和内在之间没有分隔。宗教生活是一种有秩序的、勤勉的、面对内心真实的、没有任何幻想的生活，因而你过着一种有秩序的、正义的生活。当这种生活被坚定地建立起来的时候，我们就能够开始探究冥想是什么了。

>> Perhaps that word did not exist in the Western world, in its present usage until about thirty years or so ago. The Eastern gurus have brought it over here. There is the Tibetan meditation, Zen meditation, the Hindu meditation, the particular meditation of a particular guru—the yoga meditation, sitting cross-legged, breathing—you know all that. All that is called meditation. We are not denigrating

the people who do all this. We are just pointing out how absurd meditation has become. The Christian world believes in contemplation, giving themselves over to the will of god, grace and so on. There is the same thing in the Asiatic world, only they use different words in Sanskrit, but it is the same thing—man seeking some kind of everlasting security, happiness, peace, and not finding it on earth, hoping that it exists somewhere or other—the desperate search for something imperishable—the search of man from time beyond measure.

也许西方社会没有冥想这个词，直到大约三十年以前才有了现在的用法。东方古鲁们将它带到这里。有中国冥想、禅的冥想、印度冥想、特殊古鲁的特殊冥想——瑜伽冥想，盘腿而坐，呼吸——你们知道所有这些。这一切都被称为冥想。我们不是在诋毁做这些的人们，我们只是指出冥想已经变得多么荒谬。基督教世界信仰默祷。在亚洲社会也是一样，只不过他们使用不同的梵文词语，但是它是相同的东西——人们在寻求某种永恒的安全、幸福、和平，在尘世上没有发现它，就希望它存在于某个地方——对不朽之物迫切搜寻——人类从无史以来就开始这么做了。

>> So we should enquire together, deeply, into what meditation is and whether there is anything sacred, holy—not the thing that thought has invented as being holy, that is not holy. What thought creates is not holy, is not sacred, because it is based on knowledge, and how can anything that thought invents, being incomplete, be sacred? But all over the world we worship that which thought has invented.

因此我们应当一起深入探究冥想是什么，以及是否存在神圣、圣洁的东西——不是思想发明的圣洁，那不是圣洁。思想制造的东西不是圣洁的，不是神圣的，因为它是基于知识，不完整的思想发明的东西怎么可能是神圣的呢？但是全世界都崇拜思想发明的东西。

>> There is no system, no practice but the clarity of perception of a mind that is free to observe, a mind which has no direction, no choice. Most systems of meditation have the problem of controlling thought. Most meditation, whether the Zen, the Hindu, the Buddhist, the Christian, or that of the latest guru, tries to control thought; through control you centralize, you bring all your energy to a particular point. That is concentration, which means that there is a controller different from the controlled. The controller is thought, memory, and that which he is controlling is still thought—which is wandering off, so there is conflict. You are sitting quietly and thought goes off; you are like a schoolboy looking out of the window and the teacher says, "Don't look out of the window, concentrate on your book." We have to learn the fact that the controller is the controlled. The controller, the thinker, the experiencer, are,

we think, different from the controlled, from the movement of thought, from the experience. But if we observe closely, the thinker is the thought. Thought has made the thinker separate from thought, who then says, "I must control." So when you see that the controller is the controlled you totally remove conflict. Conflict exists only when there is the division. Where there is the division between the observer, the one who witnesses, the one who experiences and that which he observes and experiences, there must be conflict. Our life is in conflict because we live with this division. But this division is fallacious, it is not real, it has become our habit, our culture, to control. We never see that the controller is the controlled.

没有体系、没有练习，只有一颗自由观察、没有方向、没有选择的心进行的清晰觉察。大多数冥想体系都存在控制思想的问题。大多数冥想，无论是禅宗的、印度教的、佛教的、基督教的，或者最新的古鲁的，都试图控制思想，通过控制来集中精神，把所有的能量带到一个特殊的点。那是专注，它意味着有一个不同于被控制对象的控制者。控制者是思想、记忆，他控制的东西仍然是思想——它总在游离，因此存在冲突。你安静地坐着而思想离开了，你像一个望着窗外的小学生，老师说："别看窗外，专心看书。"我们必须了解这个事实，控制者就是被控制者。我们认为控制者、思想者、经验者不同于被控制者、思想活动和经验。但是如果我们密切观察，思想者就是思想。思想使得思想者和思想分离，他说"我必须控制"。因此当你看到控制者就是被控制者的时候，你就完全去除了冲突。冲突只存在于分裂的时候。观察者——那个看见和感受过一些东西的人——和他观察、经验的东西之间的分裂，必定会带来冲突。我们的生活处于冲突之中，因为我们伴随着这种分裂。但是这种分裂是虚妄的，不是真实的，控制已经成了我们的习惯和文化。我们从来没有看到，控制者就是被控制者。

>> So when one realizes that fact—not verbally, not idealistically, not as a Utopian state for which you have to struggle, actually in one's life that the controller is the controlled, the thinker is the thought—then the whole pattern of one's thinking undergoes a radical change and there is no conflict. That change is absolutely necessary if one is meditating because meditation demands a mind that is highly compassionate, and therefore highly intelligent, with an intelligence which is born out of love, not out of cunning thought. Meditation means the establishment of order in one's daily life, so that there is no contradiction; it means having rejected totally all the systems of meditation so that one's mind is completely free, without direction; so that one's mind is completely silent. Is that possible? Because one is chattering endlessly; the moment one leaves this place, one will start chattering. one's mind will continue everlastingly occupied, chattering, thinking, struggling, and so there is no space. Space is necessary to have silence, for a mind that is practising, struggling, to be silent is never silent.

But when it sees that silence is absolutely necessary—not the silence projected by thought, not the silence between two notes, between two noises, between two wars, but the silence of order—then in that silence, truth, which has no path to it, exists. Truth that is timeless, sacred, incorruptible. That is meditation.

20th September, 1981

因此当你认识到那个事实，不是在口头上、理想上认识到，不是出于一种乌托邦的状态——那样你就不得不挣扎——而是在你的生活中认识到，控制者就是被控制者，思想者就是思想，那么你思想的整个模式就会经历一个根本性的转变，其中没有冲突。如果你在冥想，那种转变是绝对必要的，因为冥想需要一颗高度慈悲且高度智慧的心，智慧出于爱，而不是出于狡猾的思想。冥想意味着日常生活中秩序的建立，因而没有矛盾，它意味着彻底拒绝所有的冥想体系，因而让你的心彻底自由，没有方向，于是你的心彻底安静下来。这可能吗？因为你喋喋不休，你离开这里的时候就会开始喋喋不休。你的心将永远持续地被喋喋不休、思考、挣扎占据，因而没有空间。对安静来说，空间是必要的，因为一颗在练习的、挣扎着想要静默的心永远不会安静下来。当它看到安静是绝对必要的时候——不是思想投射的安静，不是两个音符、两个声音、两次战争之间的安静，而是有秩序的安静——在那个静默中，真理这个无路可寻的东西就会存在，无始无终的、神圣的、不朽的真理。那就是冥想。

一九八一年九月二十日

北京市版权局著作权合同登记号　图字：01-2020-2611

图书在版编目（CIP）数据

狡猾的思想 /（印）克里希那穆提著；张春城译 .— 北京：北京时代华文书局，2022.9

书名原文：The Network of Thought

ISBN 978-7-5699-3773-2

Ⅰ.①狡…　Ⅱ.①克… ②张…　Ⅲ.①人生哲学–通俗读物　Ⅳ.① B821-49

中国版本图书馆 CIP 数据核字 (2020) 第 108961 号

狡猾的思想

JIAOHUA DE SIXIANG

著　　者 | ［印］克里希那穆提
译　　者 | 张春城

出 版 人 | 陈　涛
选题策划 | 刘昭远
责任编辑 | 周海燕
执行编辑 | 胡元曜
责任校对 | 陈冬梅
装帧设计 | 柒拾叁号
责任印制 | 訾　敬

出版发行 | 北京时代华文书局 http://www.bjsdsj.com.cn
北京市东城区安定门外大街 136 号皇城国际大厦 A 座 8 层
邮编：100011　电话：010 - 64263661　64261528
印　　刷 | 北京盛通印刷股份有限公司　010 - 83670070
(如发现印装质量问题，请与印刷厂联系调换)
开　　本 | 787 mm × 1092 mm　1/32　印　张 | 10　字　数 | 150 千字
成品尺寸 | 185 mm × 125 mm
版　　次 | 2022 年 9 月第 1 版　印　次 | 2022 年 9 月第 1 次印刷

定　　价 | 49.80 元